Math 123 Finite Mathematics

Note-Taking Guides

May 2026

AF754620

Cengage

Australia • Brazil • Canada • Mexico • Singapore • United Kingdom • United States

Cengage

Math 123 Finite Mathematics, Note-Taking Guides

Math 123 Finite Mathematics: Note-Taking Guides
Angela Gentry, Jack Ryan, April Shilling-Lambert, Kim Smith, Tracy Cook and Shel Swenson

WCN: 01-100-496

ISBN: 979-8-214-53478-7

Cengage
10650 Toebben Drive, Suite A
Independence, KY 41051
USA

Math 123 - Finite Mathematics

Note-Taking Guides

Table of Contents

Math 123 Formula Sheet

Formula A. $I = Prt$

Formula B. $A = P(1 + rt)$

Formula C. $A = P(1+i)^n, \ i = \dfrac{r}{m}, \ n = mt$

Formula D. $r_{\text{eff}} = \left(1 + \dfrac{r}{m}\right)^m - 1$

Formula E. $r_{\text{eff}} = \left(\dfrac{A}{P}\right)^{1/t} - 1$

Formula F. $S = R\left[\dfrac{(1+i)^n - 1}{i}\right]$

Formula G. $P = R\left[\dfrac{1-(1+i)^{-n}}{i}\right]$

Formula H. $A^{-1} = \dfrac{1}{ad-bc}\begin{bmatrix} d & -b \\ -c & a \end{bmatrix}$, if $ad - bc \neq 0$

Formula I. $p_1 = \dfrac{d-c}{a+d-b-c}$

Formula J. $q_1 = \dfrac{d-b}{a+d-b-c}$

Formula K. $E = \dfrac{ad-bc}{a+d-b-c}$

Formula L. $E = PAQ$

Formula M. $E = \dfrac{1}{Z} - k$

Unit 1

Mathematics of Finance

This unit focuses on applying mathematical techniques to solve real-world financial problems involving interest, annuities, loans, and investments. Students learn to model and analyze financial situations using algebraic formulas and logical reasoning.

NAME: ______________________ EMAIL: ____________________ @vols.utk.edu

Lesson 1: Simple & Compound Interest

Learning Objectives:

I. Solve a financial problem involving simple interest.

II. Solve a financial problem involving compound interest.

III. Solve a financial problem involving the effective rate of interest.

I. Simple Interest

Simple interest is interest that is computed on the original principal only. If I is the interest on a principal P (in dollars) at an interest rate of r per year for t years, then we have:

Formula A. $I = Prt$

The **accumulated amount** A is the sum of the principal and interest after t years, given by:

Formula B. $A = P(1 + rt)$

For Examples 1 - 3: Write the formula that is needed, fill in the formula with the corresponding numbers, then solve. Show all work. Round the final answer to 2 decimal places, and include units with the final answer.

Example 1 (Simple interest on \$500 deposit)

(a.) Find the simple interest on a \$500 investment made for 2 years at an interest rate of 8% per year.

(b.) What is the accumulated amount?

Example 2 (Simple interest on bank deposit that grew to \$3,100)

A bank deposit paying simple interest at a rate of 5% per year grew to a sum of \$3,100 in 10 months.

(a.) Find the principal.

(b.) What is the interest earned in that period?

Example 3 (Joe's T-Bills)

Joe purchased \$10,000 worth of 26-week T-Bills for \$9,750. What will be the rate of return on his investment?

II. Compound Interest

The **conversion period** is the interval of time between successive interest calculations. If interest at a nominal rate of r per year is compounded m times per year on a principal of P dollars, then the simple interest rate per conversion period is:

Formula $$i = \frac{r}{m} \quad \left(\frac{\text{annual interest rate}}{\text{periods per year}}\right)$$

Compound interest is interest that is periodically added to the principal and therefore earns interest at the same rate (you earn interest on both the principal and the previous interest earned).

Formula C. $$A = P(1+i)^n, \quad i = \frac{r}{m}, \quad n = mt$$

i = simple interest rate per conversion period
A = accumulated amount at the end of n conversion periods (or Future Value FV)
P = principal (or Present Value PV)
r = nominal interest rate per year
m = number of conversion periods per year
t = term (in years)

For Examples 4 - 7: Write the formula that is needed, fill in the formula with the corresponding numbers, then solve. Show all work. Round the final answer to 2 decimal places, and include units with the final answer.

Example 4 (Accumulated amount on $2,000 deposit - simple interest)

Find the accumulated amount after 4 years if $2,000 is invested in an account earning simple interest at a rate of 7.5% per year.

Example 5 (Accumulated amount on $2,000 deposit - compound interest)

Find the accumulated amount after 4 years if $2,000 is invested in an account earning interest at a rate of 7.5% per year compounded: (a.) annually, (b.) semiannually, and (c.) monthly.

Example 6 (Present value of $6,500 due in 5 years)

Find the present value of $6,500 due in 5 years at a rate of 3% per year compounded quarterly.

Example 7 (Time to double $5,000)

How long will it take an investment of $5,000 to double if the investment earns interest at a rate of 5% per year compounded monthly?

III. Effective Rate of Interest

The **effective rate of interest** is the annual rate of interest that, when compounded annually, will yield the same accumulated amount as the nominal rate compounded m times per year (over the same term). The effective rate of interest is also called the **annual percentage yield (APY)**.

Formula D. $r_{\text{eff}} = \left(1 + \frac{r}{m}\right)^m - 1$

r_{eff} = effective rate of interest
r = nominal interest rate per year
m = number of conversion periods per year

Formula E. $r_{\text{eff}} = \left(\frac{A}{P}\right)^{1/t} - 1$

r_{eff} = effective rate of interest
A = accumulated amount after t years
P = principal (original investment)

For Examples 8 - 9: Write the formula that is needed, fill in the formula with the corresponding numbers, then solve. Show all work. Round the final answer to 2 decimal places, and include units with the final answer.

Example 8 (Georgia's annual rate of return)

Georgia purchased a home in September 2008 for \$300,000. In September 2014, she sold the house and made a net profit of \$66,000. Find the effective annual rate of return on her investment over the 6-year period.

Example 9 (Bendix Mutual Fund vs. Acme Mutual Fund)

In the last 5 years, Bendix Mutual Fund grew at a rate of 6.4% per year compounded quarterly. Over the same period, Acme Mutual Fund grew at a rate of 6.5% per year compounded semiannually. Which mutual fund has the better rate of return?

Additional Notes

NAME: ______________________ EMAIL: ________________ @vols.utk.edu

Lesson 2: Annuity Accounts - Accumulation Phase

Learning Objectives:

I. Solve a financial problem involving a future value annuity (i.e., the accumulation phase).

II. Identify a sinking fund and solve the corresponding future value annuity problem.

I. Accumulation Phase

So far, we have looked at situations where a fixed amount of money earns interest over a period of time. However, investments like retirement funds also involve making periodic deposits into an account earning interest. An **annuity account** is an account earning interest into which you make periodic deposits or from which you make periodic withdrawals. **Annuity** refers to the sequence of deposits or withdrawals. The time period in which these payments are made is called the **term** of the annuity.

We will study annuities that meet the following criteria:

1. The terms are given by fixed intervals (annuity certain).
2. The periodic payments/withdrawals are equal in size (for example: mortgage payments).
3. The payments/withdrawals are made at the end of the payment period (ordinary annuity).
4. The payment periods coincide with the interest conversion period (simple annuity).

The period during which you are contributing to your annuity account is the **accumulation phase**, during which the value of the account increases.

Future Value of Annuity Formula

Formula F. $$S = R\left[\frac{(1+i)^n - 1}{i}\right]$$

$S =$ Future Value (FV)
$R =$ payment amount (PMT)
$n =$ number of payments
$i =$ interest per period

Example 1 (Recurring deposits)

At the end of every month for the next 6 years, you will deposit \$500 into an account that earns interest at a rate of 2.9% per year compounded monthly. How much money will be in the account at the end of 6 years? (Round the final answer to two decimal places.)

(a.) Solve using the appropriate formula: write the formula that is needed, fill in your formula with the corresponding numbers, and solve.

(b.) Solve using a TVM solver. Fill in the information that is typed into the solver.

N = *(Periods)*			PMT = *(Payments)*	
I% = *(Annual Rate)*			FV = *(Future Value)*	
PV = *(Present Value)*			P/Y and C/Y = *(Compounding)*	

Final Answer:

For Examples 2 - 3: Write the formula that is needed, fill in the formula with the corresponding numbers, then solve. Show all work. Round the final answer to 2 decimal places, and include units with the final answer.

Example 2 (Initial & recurring deposits)

Lauren plans to deposit \$5,000 into a bank account at the beginning of next month and \$200 per month into the same account at the end of that month and at the end of each subsequent month for the next 5 years. If her bank pays interest at a rate of 3% per year compounded monthly, how much will Lauren have in the account at the end of 5 years? (Assume that she makes no withdrawals during the 5-year period.)

Example 3 (Recurring deposits that stop)

Suppose that Jacob deposits \$3,000 per year for 10 years into a Roth IRA earning interest at a rate of 5% per year compounded annually. During the next 10 years, he makes no withdrawals and no further contributions, but the account continues to earn interest at the same rate. How much will Jacob have in his retirement account at the end of the 20-year period?

II. Sinking Fund

A **sinking fund** is an account that is set up for a specific purpose at some future date. The sinking fund is an amount that needs to be accumulated by a specific date in the future; therefore, the value of the account is increasing.

Example 4 (Sinking fund)

A city has \$2.5 million worth of school bonds that are due in 20 years and has established a sinking fund to retire this debt. If the fund earns interest at a rate of 4% per year compounded annually, what amount must be deposited annually into this fund? (Round the final answer to two decimal places.)

(a.) Solve using the appropriate formula: write the formula that is needed, fill in your formula with the corresponding numbers, and solve.

(b.) Solve using a TVM solver. Fill in the information that is typed into the solver.

N = *(Periods)*			PMT = *(Payments)*	
I% = *(Annual Rate)*			FV = *(Future Value)*	
PV = *(Present Value)*			P/Y and C/Y = *(Compounding)*	

Final Answer:

Example 5 (Larry & Ben's retirement accounts)

Write the formula that is needed, fill in the formula with the corresponding numbers, then solve. Show all work. Round the final answer to 2 decimal places, and include units with the final answer.

Since he was 22 years old, Ben has been depositing \$200 at the end of each month into a tax-free retirement account earning interest at a rate of 3.5% per year compounded monthly. Larry, who is the same age as Ben, decided to open a tax-free retirement account 5 years after Ben opened his. If Larry's account earns interest at the same rate as Ben's, determine how much Larry should deposit each month into his account so that both men will have the same amount of money in their accounts at age 65.

Additional Notes

NAME: ______________________ EMAIL: ____________________ @vols.utk.edu

Lesson 3: Annuity Accounts - Annuitization Phase

Learning Objectives:

I. Solve a financial problem involving a present value annuity (i.e., the annuitization phase).

II. Identify a loan with recurring payments, solve the corresponding present value annuity problem, or create an amortization table.

I. Annuitization Phase

The period during which you are making withdrawals from your annuity account is the **annuitization phase** (or **payout phase**), during which the value of the account decreases.

Present Value of Annuity Formula

Formula G. $$P = R\left[\frac{1-(1+i)^{-n}}{i}\right]$$

P = Present Value (PV)

R = payment amount (PMT)

n = number of payments

i = interest per period

Example 1 (Periodic withdrawals)

You want to establish an account from which you can withdraw \$800 every quarter for 7 years. The account will earn interest at a rate of 6% per year compounded quarterly. How large must the account be? (Round the final answer to two decimal places.)

(a.) Solve using the appropriate formula: write the formula that is needed, fill in your formula with the corresponding numbers, and solve.

(b.) Solve using a TVM solver. Fill in the information that is typed into the solver.

N = *(Periods)*			PMT = *(Payments)*	
I% = *(Annual Rate)*			FV = *(Future Value)*	
PV = *(Present Value)*			P/Y and C/Y = *(Compounding)*	

Final Answer:

Example 2 (State lottery commission)

Write the formula that is needed, fill in the formula with the corresponding numbers, then solve. Show all work. Round the final answer to 2 decimal places, and include units with the final answer.

A state lottery commission pays the winner of the Million Dollar lottery 20 installments of \$50,000 per year. The commission makes the first payment of \$50,000 immediately and the other $n = 19$ payments at the end of each of the next 19 years. Determine how much money the commission must have in the bank initially to guarantee the payments, assuming that the balance on the deposit with the bank earns interest at a rate of 4% per year compounded yearly.

II. Amortization

In a typical installment loan, such as a car loan or home mortgage, we borrow an amount of money and then pay it back with interest by making fixed payments. The process of paying off a loan is called **amortizing** the loan.

Note: From the point of view of the lender, a loan is an annuitizing annuity. The lender "deposits" the amount of the loan and makes periodic "withdrawals" from the borrower (i.e. the loan payments) until the balance is zero. Loan calculations are identical to annuity calculations.

For Examples 3 & 4: Write the formula that is needed, fill in the formula with the corresponding numbers, then solve. Show all work. Round the final answer to 2 decimal places, and include units with the final answer.

Example 3 (Car lease)

The Browns have leased a car for 2 years with a payment of \$450 per month. If the money is worth 3.5% per year compounded monthly, what is the equivalent cash payment of this annuity?

Example 4 (Installment loan)

A sum of $15,000 is to be repaid over a 5 year period through equal installments made at the end of each year, with an interest rate of 7% per year compounded annually charged on the unpaid balance. Interest calculations are made at the end of each year.

(a.) Determine the size of each installment so that the loan (principal and interest) is amortized at the end of 5 years.

(b.) Verify by displaying the amortization schedule.

End of Period	Interest Charged	Repayment Made	Payment Toward Principal	Outstanding Principal
0				
1				
2				
3				
4				
5				

(c.) Find the total amount paid.

Example 5 (Car payment)

Solve using a TVM solver. Fill in the information that is typed into the solver. (Round the final answer to two decimal places.)

The price of a new car is \$40,000. An individual makes a down payment of 10% toward the purchase of the car and secures financing for the balance at a rate of 3.9% per year compounded monthly.

(a.) What monthly payment will they be required to make if the car is financed over a period of 72 months?

N = *(Periods)*			PMT = *(Payments)*	
I% = *(Annual Rate)*			FV = *(Future Value)*	
PV = *(Present Value)*			P/Y and C/Y = *(Compounding)*	

Final Answer:

(b.) What will the interest charges be with the 72-month plan?

For Examples 6 & 7: Write the formula that is needed, fill in the formula with the corresponding numbers, then solve. Show all work. Round the final answer to 2 decimal places, and include units with the final answer.

Example 6 (Joe's college loan)

Joe secured a loan of $12,000 three years ago from a bank for use toward his college expenses. The bank charged interest at a rate of 4% per year compounded monthly on his loan. Now that he has graduated from college, Joe wishes to repay the loan by amortizing it through monthly payments over 10 years at the same interest rate. Find the size of the monthly payments he will be required to make.

Example 7 (Jennifer's retirement account)

Jennifer is the owner of a video game and entertainment software retail store. She is currently planning to retire in 30 years and wishes to withdraw $10,000 per month for 20 years from her retirement account starting at that time. How much must she contribute each month for 30 years into a retirement account earning interest at a rate of 4% per year compounded monthly to meet her retirement goal?

Additional Notes

NAME: ______________________ EMAIL: ________________ @vols.utk.edu

Lesson 4: Home Loans

Learning Objectives:

I. Calculate the outstanding principal of a loan with recurring payments.

II. Calculate the balloon payment or equity of a loan with recurring payments.

III. Solve an adjustable-rate mortgage (i.e., ARM-loan) problem.

I. Outstanding Principal

The **outstanding principal** is the remaining balance of a loan that has not yet been paid (i.e. how much one still owes on the loan).

Example 1 (Outstanding principal)

Write the formula that is needed, fill in the formula with the corresponding numbers, then solve. Show all work. Round the final answer to 2 decimal places, and include units with the final answer.

A couple has purchased a house. They financed \$200,000 with interest charged at a rate of 4% per year compounded monthly.

(a.) Determine the size of each monthly installment so that the loan is amortized at the end of 30 years.

(b.) Find the total amount paid over the life of the 30-year loan.

(c.) Fill in the blanks to complete the portion of the amortization schedule.

End of Period	Interest Charged	Repayment Made	Payment Toward Principal	Outstanding Principal
119	$528.08	$954.83	$426.75	$157,996.00
120				
121	$525.23	$954.83	$429.60	$157,138.22

(d.) Can we find the outstanding principal at the end of 10 years without creating the table? If so, then find the value. If not, explain why.

(e.) Find the outstanding principal at the end of 10 years using a TVM solver. Fill in the information that is typed into the solver.

N = *(Periods)*			PMT = *(Payments)*	
I% = *(Annual Rate)*			FV = *(Future Value)*	
PV = *(Present Value)*			P/Y and C/Y = *(Compounding)*	

Final Answer:

II. Balloon Payments and Equity

A **balloon payment** is a repayment of the outstanding principal, made at the end of a loan period.

Example 2 (Balloon payment)

Solve using a TVM solver. Fill in the information that is typed into the solver.

Ally plans to secure a 5-year balloon mortgage of $129,000 toward the purchase of a condominium. Her monthly payment for the 5 years is calculated on the basis of a 30-year conventional mortgage at a rate of 2.9% per year compounded monthly.

(a.) What monthly payment will Ally be required to make for the first 5 years?

N = *(Periods)*			PMT = *(Payments)*	
I% = *(Annual Rate)*			FV = *(Future Value)*	
PV = *(Present Value)*			P/Y and C/Y = *(Compounding)*	

Final Answer:

(b.) What will Ally's balloon payment be?

N = *(Periods)*			PMT = *(Payments)*	
I% = *(Annual Rate)*			FV = *(Future Value)*	
PV = *(Present Value)*			P/Y and C/Y = *(Compounding)*	

Final Answer:

Home equity is the difference between the home's value and the outstanding balance on the property.

Example 3 (Home equity)

Solve using a TVM solver. Fill in the information that is typed into the solver.

The Smiths have purchased a \$175,000 house. They made an initial down payment of \$10,000 and secured a mortgage with interest charged at a rate of 5% per year on the unpaid balance. Interest computations are made at the end of each month.

(a.) If the loan is to be amortized over 30 years, what monthly payment will the Smiths be required to make?

N = *(Periods)*			PMT = *(Payments)*	
I% = *(Annual Rate)*			FV = *(Future Value)*	
PV = *(Present Value)*			P/Y and C/Y = *(Compounding)*	

Final Answer:

(b.) What is their outstanding principal after 10 years?

N = *(Periods)*			PMT = *(Payments)*	
I% = *(Annual Rate)*			FV = *(Future Value)*	
PV = *(Present Value)*			P/Y and C/Y = *(Compounding)*	

Final Answer:

(c.) What is their equity after 10 years?

III. ARM-Loans

An **ARM-loan** is a home loan in which the interest rate is changed periodically based on a financial index. For example, a 5/1 ARM-loan has an initial rate for the first 5 years and thereafter is adjusted every year for the remaining term of the loan. When the rate is adjusted, the new payments on the loan are found using the outstanding principal.

Example 4 (ARM-loan)

Solve using a TVM solver. Fill in the information that is typed into the solver.

Sam secured an ARM-loan to help finance the purchase of his home 5 years ago. The amount of the loan was $300,000 for a term of 30 years, with interest charged at a rate of 6% per year compounded monthly. Currently, the interest rate for his ARM-loan is 4.5% per year compounded monthly, and Sam's monthly payments are due to be reset. What will be the new monthly payment? (Round the final answer to two decimal places.)

Original Payments (6% Interest):

N = *(Periods)*			PMT = *(Payments)*	
I% = *(Annual Rate)*			FV = *(Future Value)*	
PV = *(Present Value)*			P/Y and C/Y = *(Compounding)*	

Final Answer:

Outstanding Principal at the Reset:

N = *(Periods)*			PMT = *(Payments)*	
I% = *(Annual Rate)*			FV = *(Future Value)*	
PV = *(Present Value)*			P/Y and C/Y = *(Compounding)*	

Final Answer:

New Payments (4.5% Interest):

N = *(Periods)*			PMT = *(Payments)*	
I% = *(Annual Rate)*			FV = *(Future Value)*	
PV = *(Present Value)*			P/Y and C/Y = *(Compounding)*	

Final Answer:

Additional Notes

Unit 1 Additional Resources

How do we know which financial formula to use?

There are several financial formulas on our Formula Sheet. Even though the formulas are provided on the exams, it's important to know when to use each formula. Here's a guide that helps us decide which formula to use.

Formulas A & B

Formulas A and B go together since they both deal with **simple interest**. Any time we see a problem that deals with simple interest, that's a big clue that we'll want to use either Formula A or Formula B.

- Use Formula A when the problem deals with the **interest only**.
  - Ex: "How much simple interest did they earn?"
- Use Formula B when the problem deals with the **accumulated amount**.
  - Ex: "How much was in the account earning simple interest at the end of the period?"

Formula C

Formula C is in its own group. It deals with **compound interest** and a **one-time deposit**.

Formulas D & E

Formulas D and E go together since they both deal with the **effective rate of interest**. Any time we see a problem that deals with an effective rate of interest, that's a big clue to use either Formula D or Formula E.

- Formulas D and E both give the effective rate of interest, so it really depends on what information we are given and what we are looking for.
- Use Formula D when the problem deals with the **nominal interest rate** and the **number of conversions per year** (the variable m).
- Use Formula E when the problem deals with the **principal**, the **accumulated amount**, and the **time**.

Formulas F & G

Formulas F and G go together since they both deal with **annuity accounts**. The main distinction here is that we are dealing with **recurring deposits or withdrawals.**

- Formula F is used when we are making **recurring deposits into the account**. This means that the value of the account is **increasing.**
- Formula G is used when we are making **recurring withdrawals out of the account**. This means that the value of the account is **decreasing.**
- We also use Formula G whenever we are dealing with **loans**. In this case, we take the perspective of the lender. The borrower making payments is like the lender getting a withdrawal from an annuity account. (We think of the borrower being the "annuity account.")

Guide to the TVM Solver

When we use the TVM solver, it is important to show work by listing the values that we plug into the TVM solver. This is demonstrated in the note-taking guides; we fill in a table to demonstrate the values that we are plugging in for each variable. We write the word "alpha" in the box of the variable we are solving for.

What does each variable stand for?

- N is the total number of payments (or, equivalently, the total number of compounding periods). $N = mt$, where m is the number of conversion periods per year and t is the time (in years). This may be labeled on the TVM solver as "Periods."
- I% is the *annual* interest rate. When using the TVM solver, leave the value in percent form (do NOT convert to decimal form) and don't include the % symbol. This may be labeled on the TVM solver as "Annual Rate" or "Rate."
- PV is the "present value." This is the amount of money in the account at the beginning.
- PMT is the "payment." This is the amount of money that is going into (or coming out of) the account on a recurring basis. (If there are no recurring deposits or withdrawals, then this value will be 0.)
- FV is the "future value." This is the amount of money in the account at the end.
- P/Y and C/Y refer to the number of payments (or conversion periods) per year. In the situations we encounter in this class, P/Y and C/Y will always be the same (because we will always deal with simple annuities). This may be labeled on the TVM solver as "Compounding."

Important Reminders

- Anything paid *out of pocket* is represented using a negative value. This is true for any variable that represents a dollar value.
- When using the TVM solver, we should have all of information except for the one variable we are trying to find. Type in all of the information that we have and then click on the appropriate button to find the value we are looking for.
- When reporting the *final answer*, always be sure to round to two decimal places, include units, and omit the negative sign (when applicable).

Guide to Amortization Tables

Amortization tables are useful because they display the exact amount of each monthly payment that goes toward principal vs. the amount that goes toward interest and the outstanding principal after each payment is made.

End of Period	Interest Charged	Repayment Made	Payment Toward Principal	Outstanding Principal
0				
1				
2				

What do the columns mean?

- **End of Period:** the number of payment periods that have passed since we began making payments on the loan. The end of period 0 refers to the start of the loan.
- **Interest Charged:** the amount of interest that is charged over the payment period.
- **Repayment Made:** the recurring payment amount.
- **Payment Toward Principal:** the amount of the payment that is put toward the principal of the loan.
- **Outstanding Principal:** the remaining balance of the loan that has not yet been paid (i.e. how much one still owes).

How do we fill in the entries?

- The row with period 0 is the beginning of the loan. We leave everything empty except for the outstanding principal (since we aren't paying anything in period 0). The outstanding principal is the original loan amount.
- To find the interest charged during a period, we take the outstanding principal from the previous period and multiply it by $i = {}^{r}/_{m}$. This gives us the dollar amount charged in interest over this period.
  (Interest Charged) = (Previous Outstanding Principal)$\cdot({}^{r}/_{m})$
- The repayment made is always the same (except in period 0 when we leave this entry blank). This is the payment amount (which is the value of R in Formula G).
- The payment toward principal is the amount of the payment that goes toward principal rather than interest. To compute this value, subtract the interest charged from the repayment made.
  (Payment Toward Principal) = (Repayment Made) − (Interest Charged)
- The outstanding principal is the principal remaining after the payment is made. To compute this value, subtract the payment toward principal from the previous row's outstanding principal.
  (Outstanding Principal) = (Previous Outstanding Principal) − (Payment Toward Principal)

Important Reminders

- Round to two decimal places in every box. Use the rounded answers from previous calculations.
- Sometimes, the outstanding principal after the final payment period will not work out to be exactly 0. That occurs since we are using the rounded monthly payment. In this case, we report the answer that we get (do not attempt to modify the entries to force the final entry to be 0).

Unit 2

Linear Systems

This unit introduces methods for solving systems of linear equations using algebraic techniques and matrix operations. It provides a foundation for applications in business, economics, computer science, and operations research, focusing on interpreting and organizing data efficiently.

NAME: ______________________ EMAIL: ____________________ @vols.utk.edu

Lesson 5: An Introduction to Systems of Linear Equations

Learning Objectives:

I. Use appropriate vocabulary to describe the possible number of solutions to a system of linear equations and classify the system.

II. Solve a system of linear equations in two variables using the graphing method.

III. Solve a system of two linear equations in two variables using the algebraic elimination method.

IV. Determine the equilibrium quantity and price for a system consisting of the supply and demand equations.

I. Solutions to a System of Equations

A **system of linear equations** is a collection of two or more linear equations in two or more variables. A **solution to a system of linear equations** is a coordinate that satisfies each equation in the system simultaneously.

How many solutions can a system of linear equations have?

-

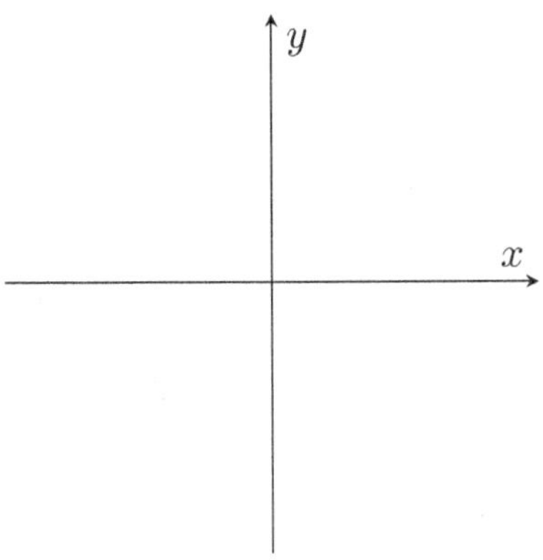

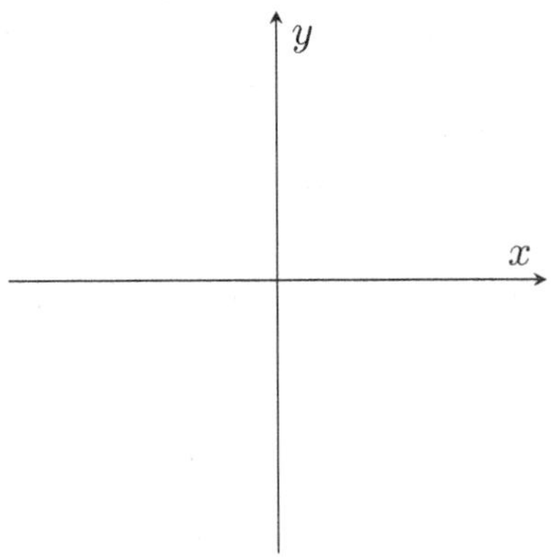

-

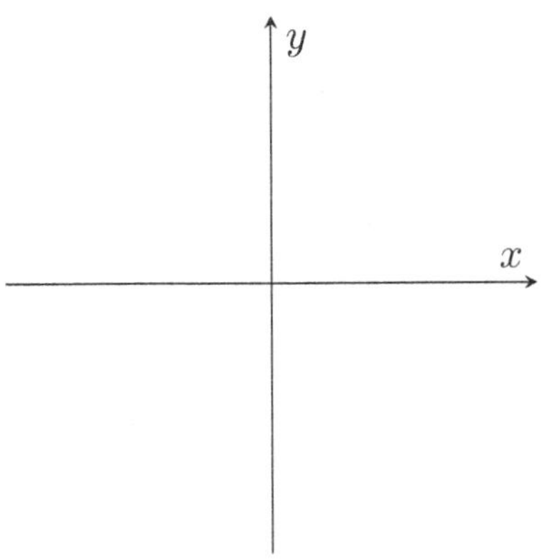

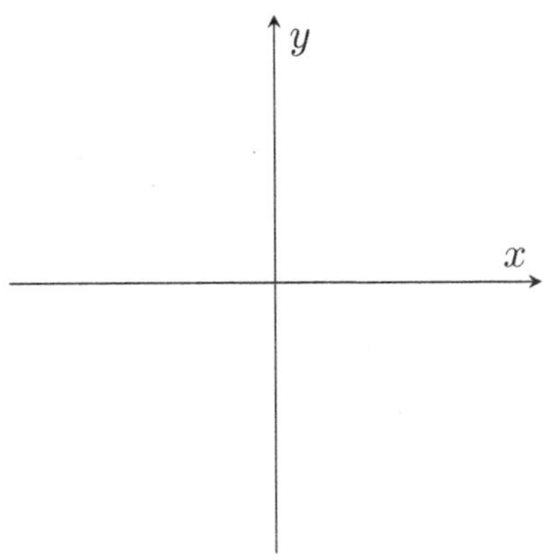

•

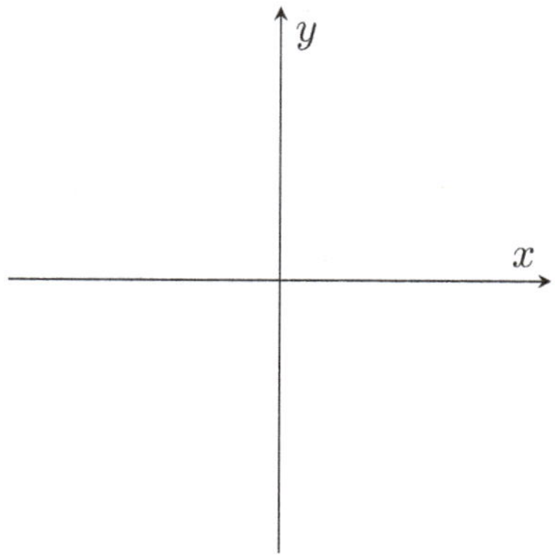

Equations in a system composed of linear equations in three variables (x, y, and z) represent a plane in three-dimensional space, and the solution(s) of the system is precisely the point(s) of intersection of the planes defined by the linear equations that make up the system. Below are illustrations of some possibilities with three equations:

Exactly 1 Solution:

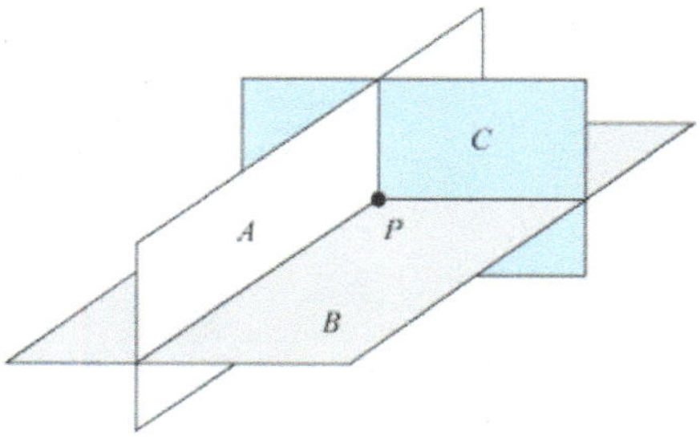

No Solution:

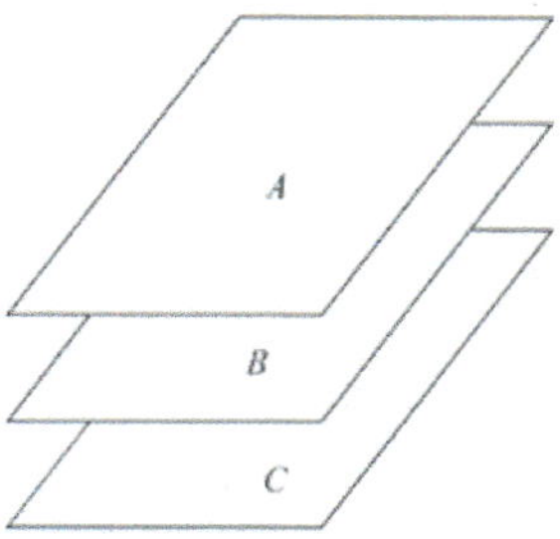

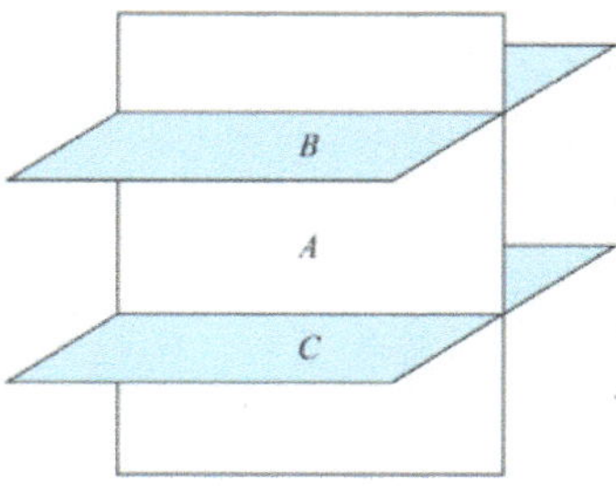

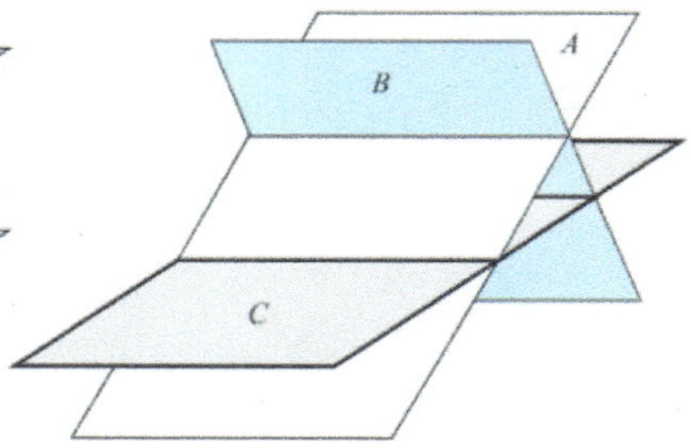

Infinitely Many Solutions:

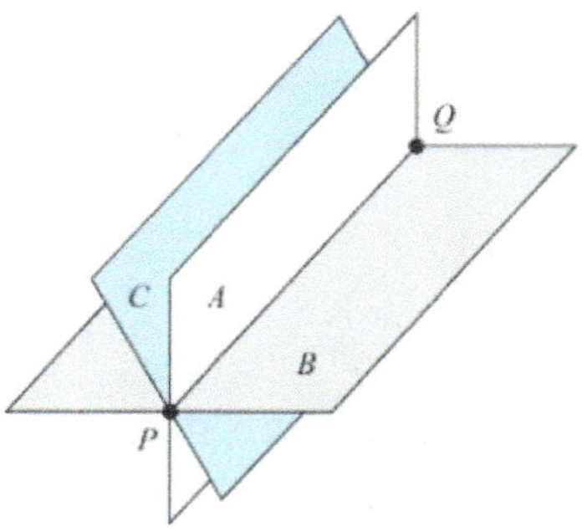

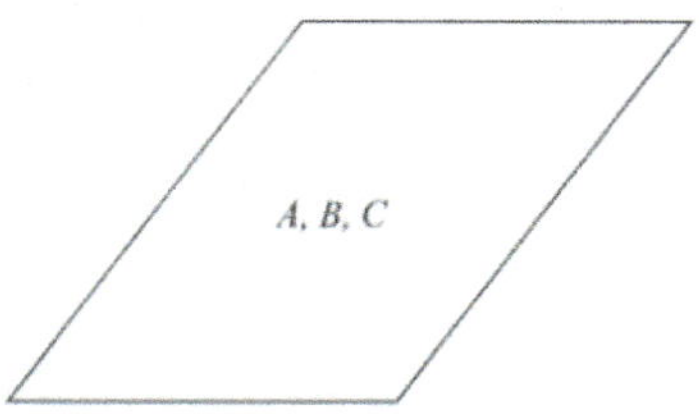

II. Solving Graphically

To solve a system of linear equations in two variables graphically:

1. Find the x- and y- intercepts for each line.
2. Adjust the scales on each axis as needed (scales are assumed to count by 1's unless otherwise specified).
3. Use the intercepts to graph the lines (and label the lines on your graph).
4. Find the solution (i.e. the point(s) where the lines intersect).

For Examples 1 - 3: Solve the system of equations graphically.

Example 1 (Solve system graphically)

$$\begin{cases} x - 2y = -5 \\ 3x + 2y = 1 \end{cases}$$

	x-intercept	y-intercept
Line 1		
Line 2		

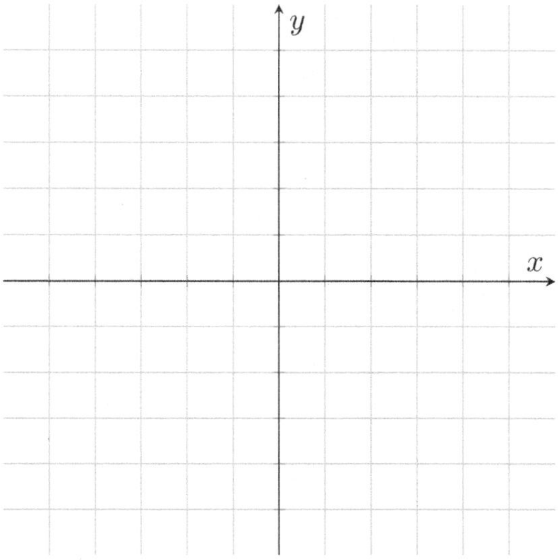

Example 2 (Solve system graphically)

$$\begin{cases} 5x - y = 8 \\ 10x - 2y = 16 \end{cases}$$

	x-intercept	y-intercept
Line 1		
Line 2		

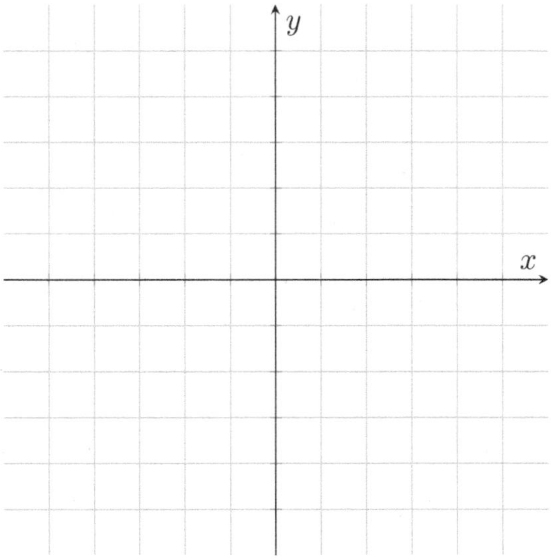

Example 3 (Solve system graphically)

$$\begin{cases} 2x + 2y = 10 \\ x + y = 4 \end{cases}$$

	x-intercept	y-intercept
Line 1		
Line 2		

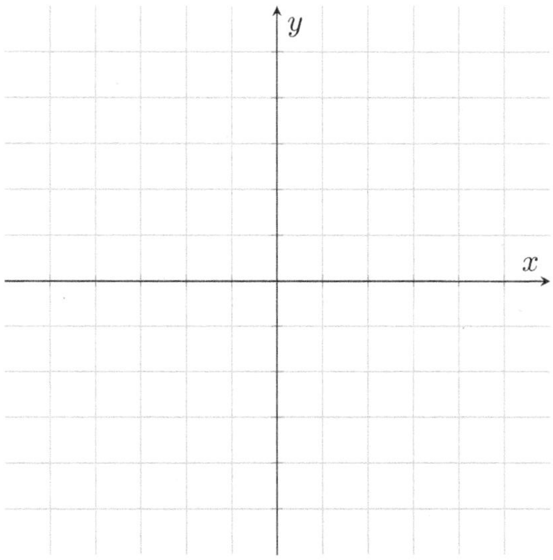

III. Solving Algebraically

To solve a system of two linear equations in two variables algebraically using the elimination method:

1. Multiply one, or both equations, by appropriate constants so that one of the variables in one equation has a coefficient c and the same variable in the other equation has a coefficient of $-c$.
2. Add the equations together to "eliminate" this variable.
3. Solve for the variable that has not been eliminated with the resulting equation.
4. Substitute the value of the variable that you just obtained into either equation to solve for the variable you eliminated earlier.

For Examples 4 - 6: Solve the system of equations using the algebraic elimination method.

Example 4 (Solve system algebraically)

$$\begin{cases} x - 2y = -5 \\ 3x + 2y = 1 \end{cases}$$

Example 5 (Solve system algebraically)

$$\begin{cases} 5x - y = 8 \\ 10x - 2y = 16 \end{cases}$$

Example 6 (Solve system algebraically)

$$\begin{cases} 2x + 2y = 10 \\ x + y = 4 \end{cases}$$

IV. Equilibrium Point

Consumer demand is the amount that consumers buy.

Example 7 (Graph consumer demand)

A bike shop held an annual sale. The consumer price and demand relationship for the special was $p = -2x + 179$, where x is the number of bikes in demand at the price p. Graph the consumer demand function.

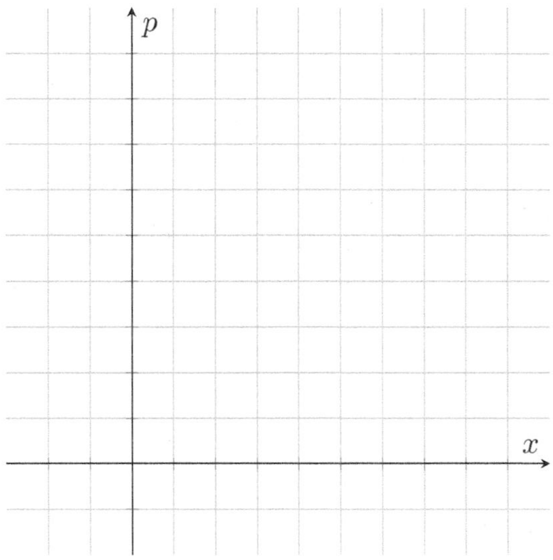

Note: A decrease in price leads to a higher demand.

Producer supply is the quantity produced by the supplier.

Example 8 (Graph producer supply)

A bike shop held an annual sale. The producer price and supply relationship for the special was $p = 1.5x + 53$, where x is the number of bikes supplied at the price p. Graph the producer supply function.

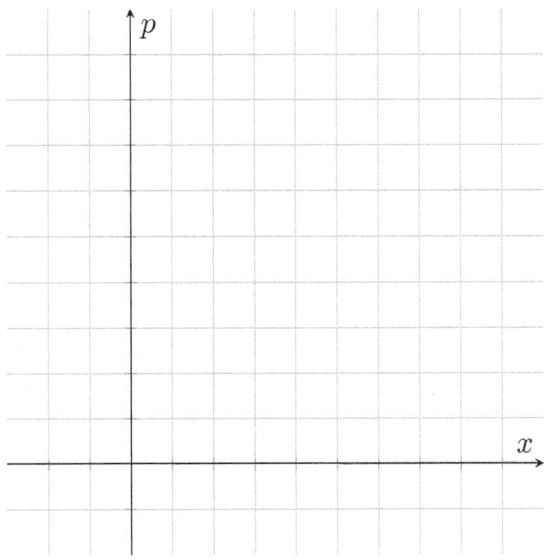

Note: An increase in price leads to a higher supply.

The demand for many products relates to the price of the product. As the price decreases, the demand by consumers increases and the amount that the producer is willing or able to supply decreases. The **equilibrium point** is the number of units and price that satisfy both the demand and supply equations; in other words, it is the solution to the system of equations. The x-value of the equilibrium point is called the **equilibrium quantity**. The p-value of the equilibrium point is called the **equilibrium price**.

Example 9 (Graph consumer demand and producer supply)

Consider the bicycle sale from the previous two examples. The consumer price and demand relationship for the special was $p = -2x + 179$ and the producer price and supply relationship for the special was $p = 1.5x + 53$, where x is the number of bikes and p is the price.

Graph the consumer demand function and producer supply function together. Find the equilibrium point.

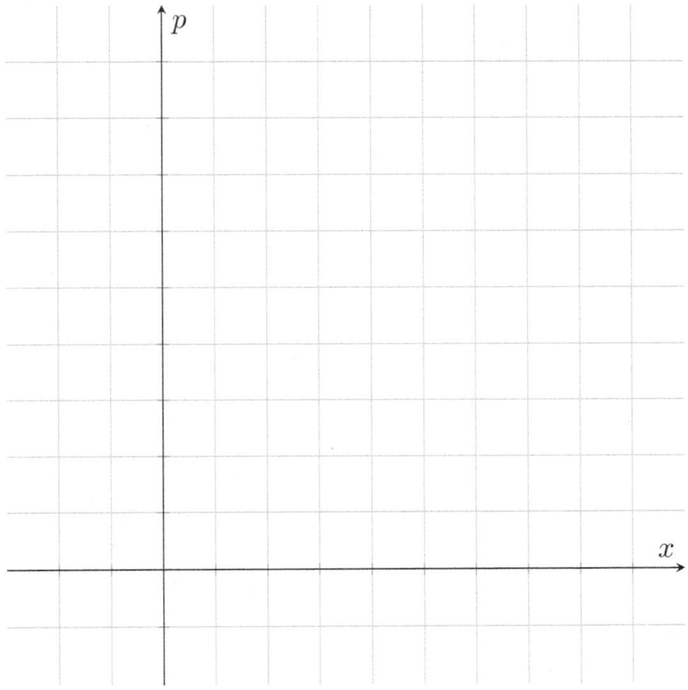

Example 10 (CD player supply and demand)

The supply and demand for a CD player in The Discount Store are given for two prices:

Demand	Supply	Price
10	40	\$130
25	10	\$100

(a.) Find the linear demand equation for the CD player.

(b.) Find the linear supply equation for the CD player.

(c.) Find the equilibrium quantity and price.

Additional Notes

NAME: ______________________ EMAIL: ____________________ @vols.utk.edu

Lesson 6: Gauss-Jordan Elimination Method

Learning Objectives:

I. Describe the size and entries of a matrix. Convert between a system of linear equations and the corresponding coefficient, constant, or augmented matrix.

II. Recognize and perform appropriate matrix row operations; pivot a matrix about a given element.

III. Determine whether a matrix is in row-reduced form. Solve a system of linear equations using the Gauss-Jordan elimination method.

IV. Given an application problem, define the variables, set up a system of equations, and solve using the Gauss-Jordan elimination method.

I. Matrices

A **matrix** (plural: matrices) is an ordered rectangular array of numbers.

A matrix with m rows and n columns has a **size (or dimension)** of ______________ .

__________ refers to the entry in row i and column j in the matrix A.

Example 1 (Size and entries of matrix)

Consider the matrix $B = \begin{bmatrix} 1 & 3 & 2 \\ -1 & 4 & 7 \end{bmatrix}$.

(a.) What is the size of the matrix B?

(b.) Find b_{12}.

(c.) Find b_{21}.

An **augmented matrix** is a combined matrix that represents a system of equations. Each row represents an equation. Each column (except for the last column) represents the coefficients of a variable. A vertical line is placed between the coefficients and the column of constant terms as a visual reminder of the equal sign in the equations.

The left side of the vertical line is called the **coefficient matrix**; the right side of the vertical line is called the **constant matrix**.

Example 2 (System of equations and augmented matrices)

Write the missing system of equations or augmented matrix.

System of Equations	Augmented Matrix
$\begin{cases} x - 2y = -5 \\ 3x + 2y = 1 \end{cases}$	
	$\left[\begin{array}{cc\|c} 2/3 & -4 & 8 \\ 3 & 0 & 15 \end{array}\right]$

Example 3 (Coefficient, constant, and augmented matrices)

For the given system of equations, find the coefficient matrix, constant matrix, and augmented matrix.

$$\begin{cases} 3x + 2y = 8 \\ 4x = y \end{cases}$$

Coefficient Matrix: Constant Matrix: Augmented Matrix:

II. Row Operations

The following are valid row operations that may be performed on any matrix:

1. Interchange any two rows:

2. Replace any row by a non-zero constant multiple of itself:

3. Replace any row by the sum of that row and a non-zero constant multiple of any other row:

For Examples 4 - 6: Perform the operation indicated beside the matrix.

Example 4 (Row operation: switch two rows)

$$\left[\begin{array}{cc|c} 0 & 4 & 2 \\ 1 & 3 & -4 \end{array}\right] \quad R_1 \leftrightarrow R_2$$

Example 5 (Row operation: multiply a row by a non-zero constant)

$$\left[\begin{array}{ccc|c} 3 & 6 & 14 & -18 \\ 4 & 2 & 5 & 7 \\ 1 & -1 & 0 & 4 \end{array}\right] \quad \tfrac{1}{3}R_1 \to R_1$$

Example 6 (Row operation: add a non-zero multiple of one row to another)

$$\left[\begin{array}{ccc|c} 1 & 3 & 2 & -4 \\ 2 & -1 & 3 & 5 \\ 4 & 6 & -2 & 3 \end{array}\right] \quad R_2 + (-2)R_1 \to R_2$$

A column in a coefficient matrix is called a **unit column** if one of the entries in the column is 1 and the other entries in the column are 0's.

The sequence of row operations that transforms a column into a unit column is called **pivoting**. The element about which a matrix is pivoted is called the **pivot element**.

Note: You can pivot on any element in the coefficient matrix, and there is more than one way to pivot which can lead to more than one correct answer.

For Examples 7 & 8: Pivot the matrix about the boxed element.

Example 7 (Pivot on row 1, column 1)

$$\left[\begin{array}{cc|c} \boxed{2} & 5 & 8 \\ 1 & 1 & 2 \end{array}\right]$$

Example 8 (Pivot on row 2, column 1)

$$\left[\begin{array}{cc|c} 3 & 2 & 6 \\ \boxed{4} & 2 & 5 \end{array}\right]$$

III. Gauss-Jordan Elimination Method

A matrix is said to be in **row-reduced** form if all of the following are true.

1. Each row consisting entirely of zeros lies below all rows having non-zero entries.
2. The first non-zero entry in each row is a 1. (This is called the **leading 1**.)
3. In any two successive non-zero rows, the leading 1 in the lower row lies to the right of the leading 1 in the upper row.
4. If a column in the coefficient matrix contains a leading 1, then the other entries in that column are zeros.

Example 9 (Row-reduced form)

Determine whether each of the following matrices are in row-reduced form. If a matrix is not in row-reduced form, state the condition(s) that is/are violated.

(a.)
$$\left[\begin{array}{ccc|c} 1 & 0 & 0 & 0 \\ 0 & 1 & 0 & 0 \\ 0 & 0 & 1 & 7 \end{array}\right]$$

(b.)
$$\left[\begin{array}{ccc|c} 1 & 1 & 0 & 5 \\ 0 & 1 & 0 & 2 \\ 0 & 0 & 0 & 0 \end{array}\right]$$

(c.)
$$\left[\begin{array}{ccc|c} 0 & 1 & 3 & 7 \\ 1 & 0 & 0 & -1 \\ 0 & 0 & 1 & 2 \end{array}\right]$$

(d.)
$$\left[\begin{array}{cc|c} 1 & 2 & 5 \\ 0 & 0 & 0 \end{array}\right]$$

Solving systems of equations using graphing or algebraic elimination works well when the number of equations and variables is small. Solving using Gauss-Jordan elimination works better for systems of any size.

The **Gauss-Jordan Elimination Method** is a process used to obtain the row-reduced form of a matrix to solve a system of equations.

1. Write the augmented matrix corresponding to the system of equations.
2. Interchange rows, if necessary, to obtain an augmented matrix in which the first entry in the first row is non-zero. Then pivot the matrix about this entry.
3. Interchange the second row with any row below it, if necessary, to obtain an augmented matrix in which the second entry in the second row is non-zero. Pivot the matrix about this entry.
4. Continue until the final matrix is in row-reduced form.

Note: Work on one column at a time, from left to right. Circle the pivot element for each augmented matrix, until you are done pivoting.

Example 10 (Solving a system using Gauss-Jordan)

Solve the system of linear equations using Gauss-Jordan elimination. Show all row operations and circle the pivot in every step.

$$\begin{cases} 2x + 3y - 6z &= -11 \\ x - 2y + 3z &= 9 \\ 3x &= 7 - y \end{cases}$$

IV. Applications

To formulate an application problem into a system of equations:

1. Define the variables: look at the question to define **specific** variables.
2. Write the equations: look at the details and constraints of the problem to write equations.

Example 11 (Theater attendance)

Define the variables and set up the system of equations. Then solve the system of equations using Gauss-Jordan elimination. Show all row operations and circle the pivot in every step.

A theater has a seating capacity of 900 and charges \$4 for children, \$6 for students, and \$8 for adults. At a certain screening with full attendance, there were half as many adults as children and students combined. The receipts totaled \$5,600. How many children, students, and adults attended the show?

Additional Notes

NAME: ______________________ EMAIL: ____________________ @vols.utk.edu

Lesson 7: Systems of Linear Equations
Underdetermined & Overdetermined Systems

Learning Objectives:

I. Analyze an augmented matrix in row-reduced form to determine when a system of equations has no solution.

II. Analyze an augmented matrix in row-reduced form to determine when a system of equations has infinite solutions and parameterize the solutions.

III. Given an augmented matrix in row-reduced form, determine whether a system of equations is overdetermined or underdetermined, and if possible, find the solution.

I. No Solutions

An **inconsistent system** is a system of equations with no solution. A **consistent system** is a system of equations that has at least one solution.

Example 1 (No Solution)

Solve the system of linear equations using the Gauss-Jordan elimination method. Show all row operations and circle the pivot in every step.

$$\begin{cases} 2x + 4y &= 10 \\ 4x + 8y &= 16 \end{cases}$$

II. Infinite Solutions

A **dependent** system is a system of equations that has infinite solutions. For a dependent system, you must write your answer using a parameter. When one variable is written in terms of another variable, we call it a **parameter**. The variables to use as parameters are those that do not correspond to unit columns with leading 1's in the row-reduced matrix.

Example 2 (Infinite Solutions)

The given augmented matrix in row-reduced form is equivalent to the augmented matrix corresponding to a system of linear equations. For each part, find the solutions.

(a.) $$\left[\begin{array}{ccc|c} 1 & 0 & 3 & -1 \\ 0 & 1 & 2 & 5 \\ 0 & 0 & 0 & 0 \end{array}\right]$$

(b.) $$\left[\begin{array}{cccc|c} 0 & 1 & 0 & 1 & 3 \\ 0 & 0 & 1 & -2 & 4 \\ 0 & 0 & 0 & 0 & 0 \\ 0 & 0 & 0 & 0 & 0 \end{array}\right]$$

III. Underdetermined & Overdetermined Systems

An **underdetermined system of equations** is a system that has *fewer* equations than variables. These systems can have no solution or infinite solutions.

An **overdetermined system of equations** is a system that has *more* equations than variables. These systems can have no solution, one solution, or infinite solutions.

Example 3 (Underdetermined vs. Overdetermined)

The given augmented matrix in row-reduced form is equivalent to the augmented matrix corresponding to a system of linear equations. For each matrix:

- Determine whether the corresponding system is underdetermined or overdetermined.
- Determine whether the system has a solution.
- Find the solution(s) if possible.

(a.) $\left[\begin{array}{ccc|c} 1 & 0 & 0 & 2 \\ 0 & 0 & 0 & 1 \end{array}\right]$

(b.) $\left[\begin{array}{ccc|c} 1 & 4 & -3 & 9 \\ 0 & 0 & 0 & 0 \end{array}\right]$

(c.) $\left[\begin{array}{cc|c} 1 & 0 & -7 \\ 0 & 1 & 15 \\ 0 & 0 & 0 \end{array}\right]$

Example 4 (Hartman Rent-A-Car)

The management of Hartman Rent-A-Car has allocated \$1,512,000 to purchase 60 new automobiles to add to their existing fleet of rental cars. The company will choose from compact, mid-sized, and full-sized cars which cost \$18,000, \$28,800, and \$39,600 each, respectively. Find formulas that express the options available to the company. Give two specific options.

Additional Notes

NAME: ______________________ EMAIL: ____________________ @vols.utk.edu

Lesson 8: Matrices

Learning Objectives:

I. Determine whether two matrices are equal, and if so, use the definition of equality to solve for variable elements.

II. Determine whether two matrices can be added or subtracted, and if possible, perform the operation.

III. Identify when a matrix is multiplied by a scalar and perform the operation.

IV. Find the transpose of a matrix.

V. Organize data from an application problem into a matrix and perform appropriate matrix operations to solve it.

I. Equality of Matrices

Two matrices are **equal** if they have the same size and their corresponding entries are equal.

Example 1 (Equality of matrices)

Solve for x, y, and z.

$$\begin{bmatrix} 7 & 9 & -8 \\ 5y+2 & -1 & z+1 \end{bmatrix} = \begin{bmatrix} 7 & 9 & -x \\ 0 & -1 & 11 \end{bmatrix}$$

II. Addition & Subtraction

If A and B are matrices of the same size, then their **sum/difference** is the matrix obtained by adding/subtracting the corresponding entries in the two matrices. If two matrices are not the same size, then they cannot be added/subtracted; in this case, we say that the sum/difference does not exist.

Example 2 (Adding and subtracting matrices)

Perform the indicated operation.

(a.) $\begin{bmatrix} 2 & -3 \\ 4 & x \\ -1 & 3 \end{bmatrix} + \begin{bmatrix} 9 & -5 \\ 0 & 13 \\ y & 4/5 \end{bmatrix}$

(b.) $\begin{bmatrix} 6 & 8 \\ 3 & -2 \end{bmatrix} - \begin{bmatrix} a & 1 \\ -5 & -3b \end{bmatrix}$

(c.) $\begin{bmatrix} 3 & -7 & 8 \\ x & -4 & 5 \end{bmatrix} + \begin{bmatrix} 1 & 0 \\ y & -3 \\ 2 & z \end{bmatrix}$

III. Scalar Multiplication

If A is a matrix and c is a real number (i.e. a scalar), then the **scalar product** cA is the matrix obtained by multiplying each entry in A by c.

Example 3 (Scalar multiplication)

Perform the indicated operation.

$$\frac{1}{3}\begin{bmatrix} 3 & -2/5 \\ y & 0 \end{bmatrix}$$

Example 4 (Matrix operations)

Solve for w, x, y, and z.

$$\begin{bmatrix} 1 & x \\ 2y & -3 \end{bmatrix} - 4\begin{bmatrix} 2 & -2 \\ 0 & 3 \end{bmatrix} = \begin{bmatrix} 3z & 10 \\ 4 & -w \end{bmatrix}$$

IV. Transpose of a Matrix

If A is an $m \times n$ matrix with elements a_{ij}, then the **transpose** of A is the $n \times m$ matrix A^T with elements a_{ji}.

Example 5 (Transpose of a matrix)

Given $C = \begin{bmatrix} 10 & 20 & 5 \\ -2 & 7 & 11 \end{bmatrix}$, find C^T.

V. Applications

Example 6 (Appliance data)

An appliance salesperson sold 15 washers, 8 dryers, and 13 microwave ovens in March. She sold 12 washers, 11 dryers, and 6 microwave ovens in April. Summarize the given information in matrix form.

Example 7 (Central bank accounts)

The matrix A represents the number of three types of bank accounts on January 1 at the Central Bank and its branches. The matrix B represents the number of accounts opened at each location during the first quarter, and the matrix C represents the number of accounts closed at each location during the same period.

$$A = \begin{array}{c} \\ \\ \\ \end{array} \begin{array}{ccc} \text{Checking accounts} & \text{Savings accounts} & \text{Fixed deposit accounts} \end{array}$$

$$A = \begin{bmatrix} 2820 & 1470 & 1120 \\ 1030 & 520 & 480 \\ 1170 & 540 & 460 \end{bmatrix} \begin{array}{l} \text{Main Office} \\ \text{Westside Branch} \\ \text{Eastside Branch} \end{array}$$

(Columns: Checking accounts, Savings accounts, Fixed deposit accounts)

$$B = \begin{bmatrix} 260 & 120 & 110 \\ 140 & 60 & 50 \\ 120 & 70 & 50 \end{bmatrix}; \qquad C = \begin{bmatrix} 120 & 80 & 80 \\ 70 & 30 & 40 \\ 60 & 20 & 40 \end{bmatrix}$$

(a.) Find a matrix D which represents the number of each type of account at the end of the first quarter at each location.

(b.) Because a new manufacturing plant is opening in the immediate area, it is anticipated that there will be a 10% increase in the number of accounts at each location during the second quarter. Write a matrix E to reflect this anticipated increase.

(c.) Find and interpret the value d_{23}.

Additional Notes

NAME: ______________________ EMAIL: ______________________ @vols.utk.edu

Lesson 9: Multiplication of Matrices

Learning Objectives:

I. Compute the dot product of a row matrix and a column matrix.

II. Determine whether two matrices can be multiplied, and if possible, perform the operation.

III. Recognize the identity matrix and its properties.

I. Dot Product

A matrix composed of a single row of numbers is called a __________ matrix.

A matrix composed of a single column of numbers is called a __________ matrix.

A matrix in which the number of rows equals the number of columns is called a __________ matrix.

The **dot product** of a row matrix and a column matrix gives a single number. To obtain the dot product, multiply the first numbers from both the row and column. Then multiply the second numbers from both, and so on. Finally, add the results together. (Note that the length of the row must match the length of the column.)

$$\begin{bmatrix} a_1 & a_2 & \dots & a_n \end{bmatrix} \begin{bmatrix} b_1 \\ b_2 \\ \vdots \\ b_n \end{bmatrix} =$$

Example 1 (Dot product)

Compute the dot product.

$$\begin{bmatrix} 4 & 0 & -1 \end{bmatrix} \begin{bmatrix} -4 \\ 1 \\ 8 \end{bmatrix}$$

Example 2 (Application of dot product)

The row matrix $\begin{bmatrix} 2.99 & 2.78 \end{bmatrix}$ represents the prices of a six-pack of soda and a package of granola bars, in that order. Let the column matrix $\begin{bmatrix} 3 \\ 4 \end{bmatrix}$ represent the quantity of six-packs of soda and granola bars purchased, in that order. Find the total cost of the purchase.

II. Product of Two Matrices

To multiply two matrices A and B together (i.e. to find $C = AB$), use the following steps:

1. Check the number of *columns* in A and the number of *rows* in B. If they are equal, then the product AB is possible. If they are not equal, then the product AB does not exist (i.e. the multiplication is not possible).
2. The number of rows in C is the same as the number of rows in A. The number of columns in C is the same as the number of columns in B.
3. To find c_{ij} (i.e. the entry in the ith row and j^{th} column of C), take the dot product of the ith row of A and the jth column of B.

Example 3 (Multiply AB and BA)

Let $A = \begin{bmatrix} 3 & -3 \\ 5 & -7 \end{bmatrix}$ and $B = \begin{bmatrix} 1 & 0 \\ 2 & -1 \end{bmatrix}$.

(a.) Compute AB.

(b.) Compute BA.

For Examples 4 - 6: Compute the product, if possible. If it is not possible to compute the product, explain why not.

Example 4 (Multiply 2×3 by 3×3)

$$\begin{bmatrix} 1 & 0 & -1 \\ 1 & 1 & 2 \end{bmatrix} \begin{bmatrix} 0 & 1 & -1 \\ 1 & 0 & 1 \\ 4 & 8 & 0 \end{bmatrix}$$

Example 5 (Multiply 2×2 by 3×2)

$$\begin{bmatrix} 5 & 0 \\ 1 & -3 \end{bmatrix} \begin{bmatrix} 1 & 0 \\ 6 & -2 \\ 4 & 3 \end{bmatrix}$$

Example 6 (Multiply 2×3 by 3×3 with variables)

$$\begin{bmatrix} 0 & -2 & 10 \\ x & 7 & 3 \end{bmatrix} \begin{bmatrix} 4 & 5 & 7 \\ y & -1 & 0 \\ 2 & u & 6 \end{bmatrix}$$

Example 7 (The Cinema Center)

The Cinema Center consists of four theaters: Cinemas I, II, III, and IV. The admission price for one feature at the Center is \$4 for children, \$6 for students, and \$8 for adults. The attendance for the Sunday matinee is given in the following matrix.

$$A = \begin{array}{c} \\ \\ \\ \\ \end{array}\begin{array}{ccc} \text{Children} & \text{Students} & \text{Adults} \end{array}$$

$$A = \begin{bmatrix} 225 & 110 & 50 \\ 75 & 180 & 225 \\ 280 & 85 & 110 \\ 0 & 250 & 225 \end{bmatrix} \begin{array}{l} \text{Cinema I} \\ \text{Cinema II} \\ \text{Cinema III} \\ \text{Cinema IV} \end{array}$$

(columns: Children, Students, Adults)

(a.) Write a column matrix B representing the admission prices.

(b.) Use matrix multiplication to find a matrix C representing the gross receipts for each theater.

(c.) Find the total revenue collected at the Cinema Center for admission that Sunday afternoon.

III. Identity Matrix

What can you multiply by the number 5 to get back 5 as an answer?

What can you add to the number 7 to get back 7 as an answer?

In general, we have no similar property for the multiplication of matrices. However, there is such a matrix for square matrices of a given size.

Example 8 (Multiplication by identity matrix)

Compute the product.

$$\begin{bmatrix} 2 & -1 & 0 \\ 5 & 3 & -1 \\ -2 & 4 & 3 \end{bmatrix} \begin{bmatrix} 1 & 0 & 0 \\ 0 & 1 & 0 \\ 0 & 0 & 1 \end{bmatrix}$$

If we let I be the $n \times n$ square matrix with ones down the main diagonal and zeros elsewhere, we have what is called the **identity matrix**. (The **main diagonal** is the diagonal that runs from the upper left to the lower right.)

If A is a square matrix of size $n \times n$ and I is the identity matrix of size $n \times n$, then ________________ .

Additional Notes

NAME: ______________________ EMAIL: ______________________ @vols.utk.edu

Lesson 10: The Inverse of a Square Matrix

Learning Objectives:

I. Recognize the inverse matrix and its properties. Determine whether two matrices are inverses of each other.

II. Find the inverse of any square matrix or describe why an inverse does not exist.

III. Convert between a system of linear equations and the corresponding matrix equation.

IV. Solve a system of linear equations using a matrix equation and the inverse of the coefficient matrix (i.e., matrix inversion).

I. Inverse Matrix

The **multiplicative inverse** of a value is a value that, when multiplied with the original number, will equal the multiplicative identity.

Find the multiplicative inverse of 2 and $\frac{3}{4}$.

The standard notation for the inverse of a is a^{-1}.

What is $\left(\frac{5}{8}\right)^{-1}$? Why?

The **inverse** of a matrix is a matrix that, when multiplied with the original matrix, will equal the identity matrix.

Let A be a square matrix of size $n \times n$. A square matrix A^{-1} of size $n \times n$ is the **inverse** of A if $AA^{-1} = A^{-1}A = I$.

- If A has an inverse, then A is called ______________________ .
- If A does not have an inverse, then A is called ______________________ .

For Examples 1 & 2: Determine whether the given matrices are inverse matrices.

Example 1 (Check if A and B are inverses)

$A = \begin{bmatrix} 1 & 2 \\ 3 & 4 \end{bmatrix}$ and $B = \begin{bmatrix} -2 & 1 \\ 3/2 & -1/2 \end{bmatrix}$.

Example 2 (Check if C and D are inverses)

$C = \begin{bmatrix} 2 & 1 \\ -2 & 2 \end{bmatrix}$ and $D = \begin{bmatrix} 1/2 & 1 \\ -1/2 & 1/2 \end{bmatrix}$.

II. How to Find the Inverse of a Matrix

We have seen that we can check whether two matrices are inverses of each other. A more challenging task is to find the inverse of a given matrix. Our process will depend on the size of the matrix.

1×1 Matrix:

The inverse of the 1×1 matrix $\begin{bmatrix} a \end{bmatrix}$ is $\begin{bmatrix} 1/a \end{bmatrix}$ since $\begin{bmatrix} a \end{bmatrix} \begin{bmatrix} 1/a \end{bmatrix} = \begin{bmatrix} 1 \end{bmatrix}$ (assuming that $a \neq 0$).

Example 3 (Inverse of 1×1 matrix)

Find the inverse of $A = \begin{bmatrix} \frac{7}{3} \end{bmatrix}$, if possible. If the inverse does not exist, explain why.

2×2 Matrix:

The inverse of the 2×2 matrix $\begin{bmatrix} a & b \\ c & d \end{bmatrix}$ is given by the following formula.

Formula H. $A^{-1} = \dfrac{1}{ad - bc} \begin{bmatrix} d & -b \\ -c & a \end{bmatrix}$, if $ad - bc \neq 0$

Note that $D = ad - bc$ is called the **determinant** of A.

Example 4 (Inverse of 2×2 matrix)

Find the inverse of $A = \begin{bmatrix} 2 & 1 \\ 5 & 1 \end{bmatrix}$, if possible. If the inverse does not exist, explain why.

$n \times n$ Matrix (for $n > 2$):

To find the inverse of the $n \times n$ matrix A (for $n > 2$):

1. Adjoin the $n \times n$ identity matrix to obtain the augmented matrix $\left[\, A \mid I \,\right]$.
2. Use the Gauss-Jordan elimination method to convert $\left[\, A \mid I \,\right]$ into a matrix of the form $\left[\, I \mid B \,\right]$, if possible. In this case, $A^{-1} = B$.
3. If you cannot obtain the form $\left[\, I \mid B \,\right]$, then A does not have an inverse (i.e. it is singular).

Example 5 (Inverse of 3×3 matrix)

Find the inverse of $A = \begin{bmatrix} 2 & 0 & 1 \\ 2 & 1 & -1 \\ 3 & 1 & -1 \end{bmatrix}$, if possible. If the inverse does not exist, explain why.

Example 6 (Inverse of 3×3 matrix)

Find the inverse of $A = \begin{bmatrix} 1 & 2 & 3 \\ 2 & 1 & 2 \\ 3 & 3 & 5 \end{bmatrix}$, if possible. If the inverse does not exist, explain why.

Note: If there is a row to the left of the vertical line in the augmented matrix containing all zeros, then the matrix does not have an inverse.

III. Matrix Representation of a System of Equations

It is often helpful to represent a system of equations as a matrix equation $AX = B$, where A is the coefficient matrix, X is the variable column matrix, and B is the constant term column matrix.

Example 7 (Write matrix equation as system of linear equations)

Write the matrix equation as a system of linear equations.

$$\begin{bmatrix} 1 & 3 & -1 \\ 2 & 0 & 4 \\ 1 & 5 & 0 \end{bmatrix} \begin{bmatrix} x \\ y \\ z \end{bmatrix} = \begin{bmatrix} -1 \\ 2 \\ 5 \end{bmatrix}$$

Example 8 (Write system of linear equations as matrix equation)

Write the system of linear equations as a matrix equation.

$$\begin{cases} x - 2y + 3z &= -1 \\ 3x + 4y - 2z - 1 &= 0 \\ 2x - 3y + 7z &= 6 \end{cases}$$

IV. Using the Inverse to Solve a System of Equations

Solve the equation $ax = b$ for x.

If $AX = B$ is the matrix equation corresponding to a system of n linear equations in n variables and A^{-1} exists,

then ________________ is the unique solution.

In Examples 9 & 10: Solve the system of equations by setting up a matrix equation and using the inverse matrix to find the solution. Write the final answer as a point.

Example 9 (Solve system using inverse matrix)

$$\begin{cases} 2x + z &= 1 \\ 2x + y - z &= 1 \\ 3x + y - z &= 1 \end{cases}$$

Example 10 (Solve system using inverse matrix)

$$\begin{cases} 2x + z &= 0 \\ 2x + y - z &= 1 \\ 3x + y - z &= 2 \end{cases}$$

Additional Notes

Graphing Lines

To graph a line, we need to know two points on that line. We plot both points and then draw the line that goes through these two points (being sure to extend the line past each point and use arrows at the ends to indicate that the line goes on forever in both directions.

To start off, we find the x- and y-intercepts. Usually, this gives us two points. We like to use these intercepts to graph since these points are usually fairly easy to find and graph. Sometimes, though, you will run into a situation where the x- and y-intercepts are both $(0, 0)$. This means that we only have *one* point. So, we need to find another point on the line in order to sketch our line.

For example, let's say that we need to graph the line given by the equation $3x + 4y = 0$. We can check that $(0, 0)$ is the x-intercept and the y-intercept. Since both intercepts occur at the same point, we need to find a second point on the line to finish graphing.

How do we find a second point?

To find a second point, we can plug any value other than 0 in for either x or y and solve for the remaining variable. Let's try $x = 1$:

$$3(1) + 4y = 0 \implies 4y = -3 \implies y = -\frac{3}{4}.$$

So, we've found that $\left(1, -\frac{3}{4}\right)$ is also on the line. But, that's not a very easy point for us to plot since it involves a fraction (it will be difficult for us to plot this point accurately on our graph).

How do we find a second point that's easy to graph?

To find a point that works better (because it's easier to plot), let's go ahead and solve for y first (before choosing a specific x-value to plug in):

$$4y = -3x \implies y = -\frac{3}{4}x$$

So, if we choose an x-value that's a multiple of 4, then the denominator in the fraction will cancel and we'll be left with a whole number for y. Let's try $x = 4$:

$$y = -\frac{3}{4}(4) = -3$$

This means that the point $(4, -3)$ is on the line - which is much easier to plot.

Empty Graphs

The following empty graphs may be useful when sketching a graph on your own paper.

Remember that axes are assumed to count by 1's unless otherwise specified. When counting by anything other than 1's, the axes need to be labeled to indicate this.

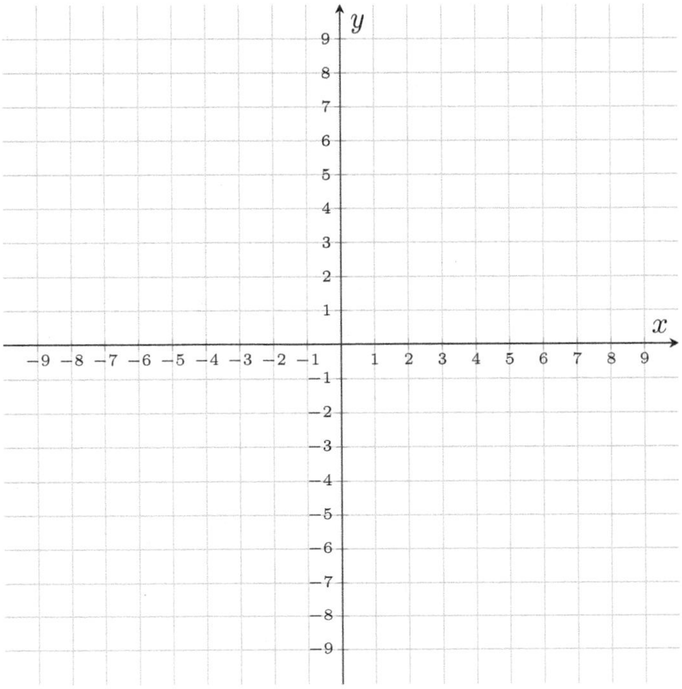

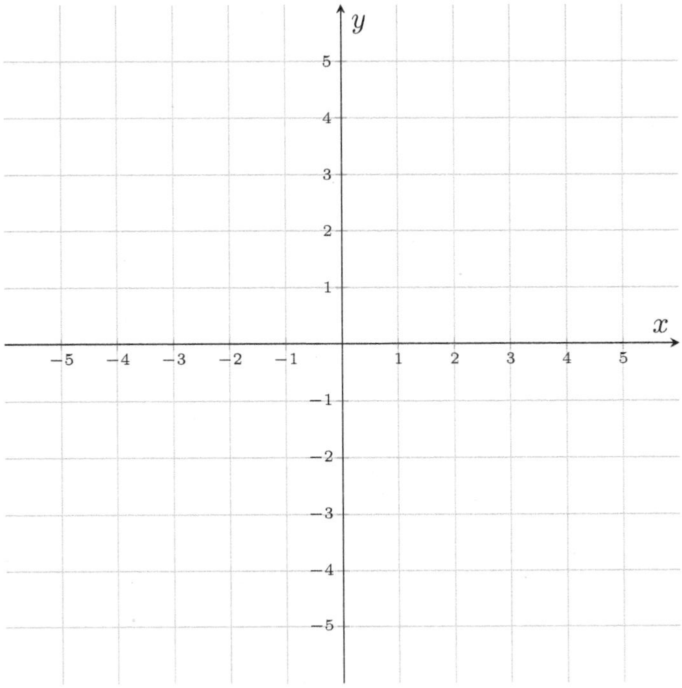

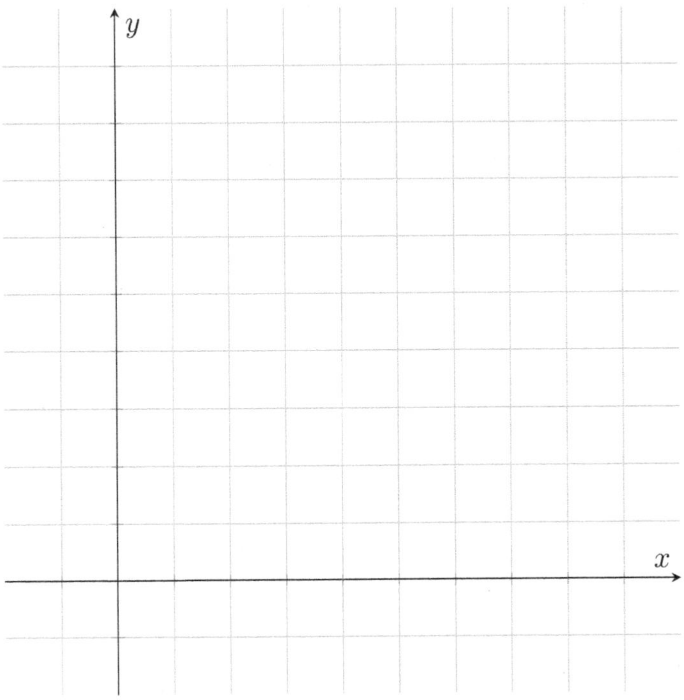

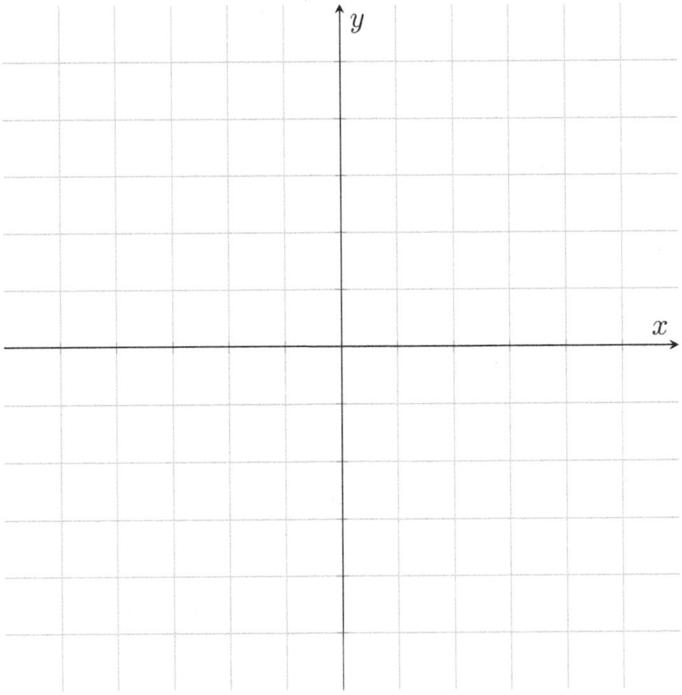

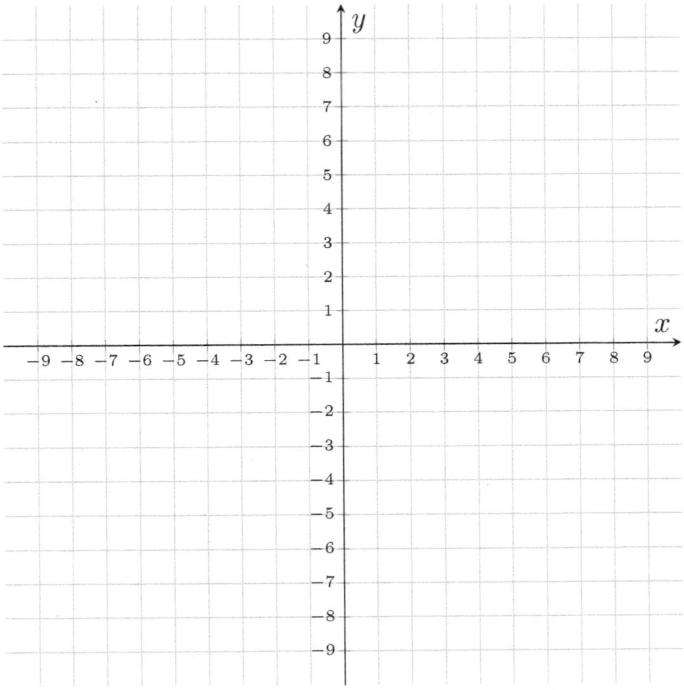
y
x
9
8
7
6
5
4
3
2
1
−1
−2
−3
−4
−5
−6
−7
−8
−9
−9 −8 −7 −6 −5 −4 −3 −2 −1 1 2 3 4 5 6 7 8 9

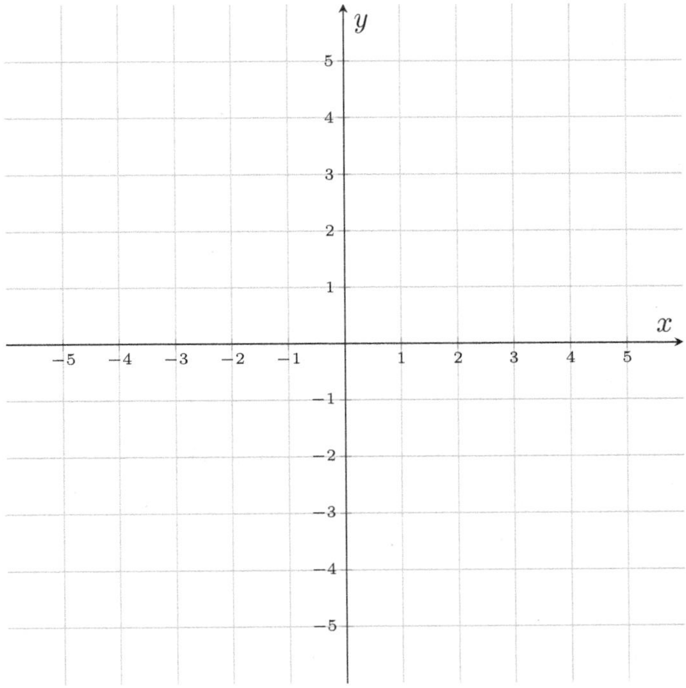
y
x
5
4
3
2
1
−1
−2
−3
−4
−5
−5 −4 −3 −2 −1 1 2 3 4 5

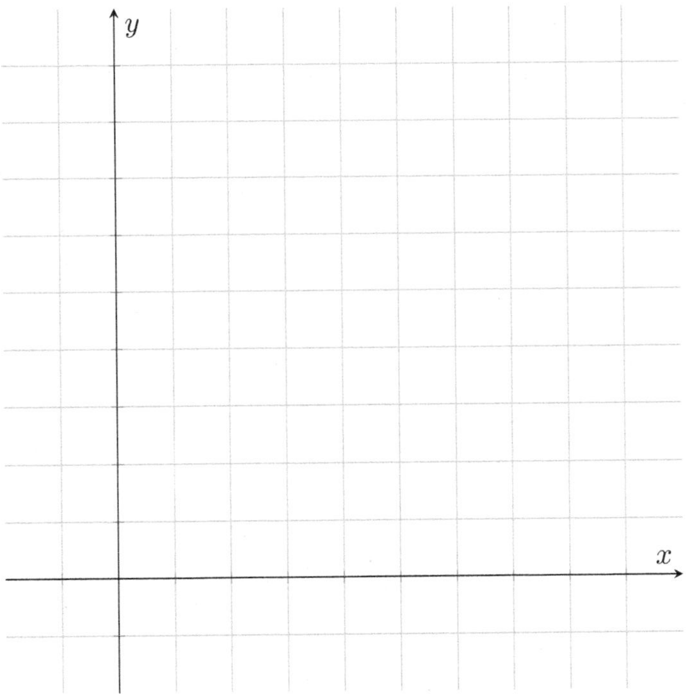
y
x

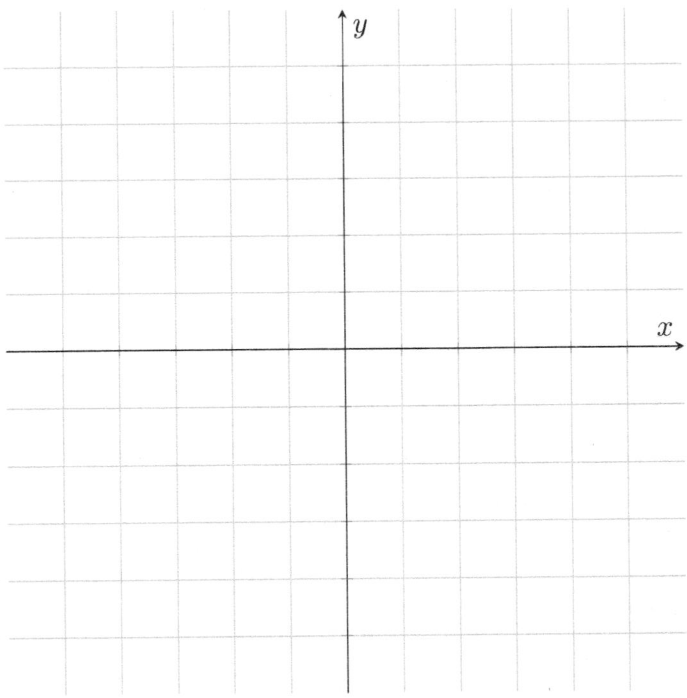
y
x

UNIT 3

LINEAR PROGRAMMING

This unit introduces linear programming (LP) as a powerful mathematical technique to find the optimal outcome (such as maximum profit or minimum cost) under a set of linear constraints. It blends algebra, graphing, and optimization to solve real-world decision-making problems in business, economics, logistics, and management.

Name: ____________________ Email: ____________________ @vols.utk.edu

Lesson 11: Graphing Systems of Linear Inequalities

Learning Objectives:

I. Sketch the graph of a linear inequality.

II. Convert between a system of linear inequalities and its graph.

III. Determine the corner points of the solution set of a system of linear inequalities.

I. Graphing Linear Inequalities

To graph a linear inequality:

1. Draw the graph of the inequality as an equation (i.e. with an equal sign).
2. Use a solid line if the inequality is $\leq$ or $\geq$, and a dashed line if the inequality is $<$ or $>$.
3. Pick any test point **not** on the line.
4. Shade (or "block out") the side that does **NOT** contain solutions. Mark an S for the side that does contain the **solution set**.

For Examples 1 - 3: Sketch the region that corresponds to the given inequality. Mark the solution with an S.

Example 1 (Graph inequality)

$6x + 4y \leq 12$

x-intercept	y-intercept	Test Point

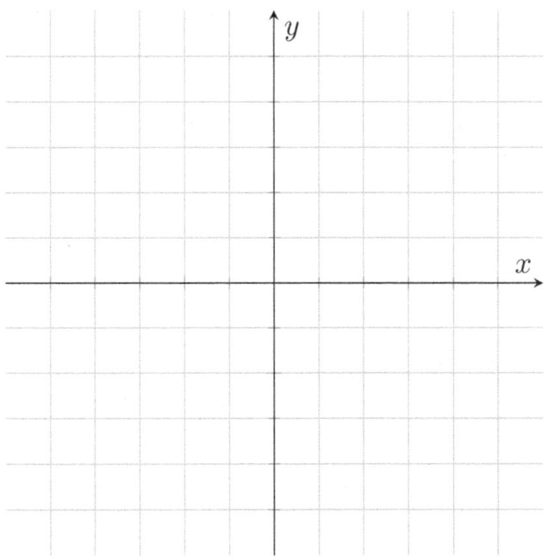

Example 2 (Graph inequality)

$6x + 4y > 12$

x-intercept	y-intercept	Test Point

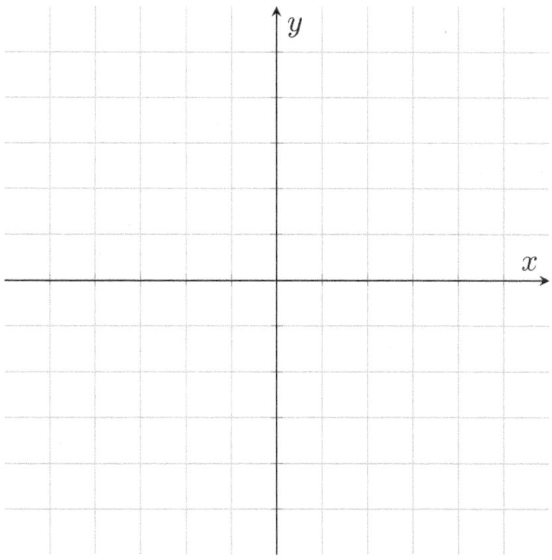

Example 3 (Graph inequality)

$x > -1$

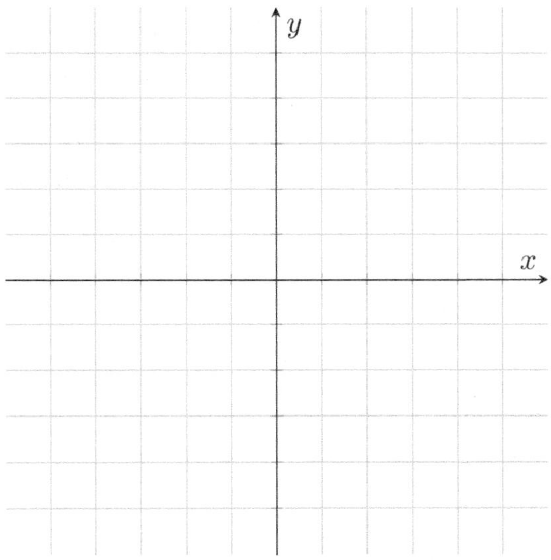

II. Graphing a System of Linear Inequalities

To graph a *system of* linear inequalities:

1. For each inequality:
   (a) Draw the graph of the inequality as an equation (i.e. with an equal sign).
   (b) Use a solid line if the inequality is $\leq$ or $\geq$, and a dashed line if the inequality is $<$ or $>$.
   (c) Pick any test point **not** on the line and shade the appropriate region (i.e. the side that is false).
2. The region left empty is the **feasible region** or (**solution set**). Mark this region with an S.

For Examples 4 - 5: Sketch the region that corresponds to the given inequalities. Label each line and mark the feasible region with an S.

Example 4 (Graph system of inequalities)

$$\begin{aligned} 2x - y &\leq 8 \\ 2x + y &> 0 \end{aligned}$$

	x-intercept	y-intercept	Test Point
Line 1			
Line 2			

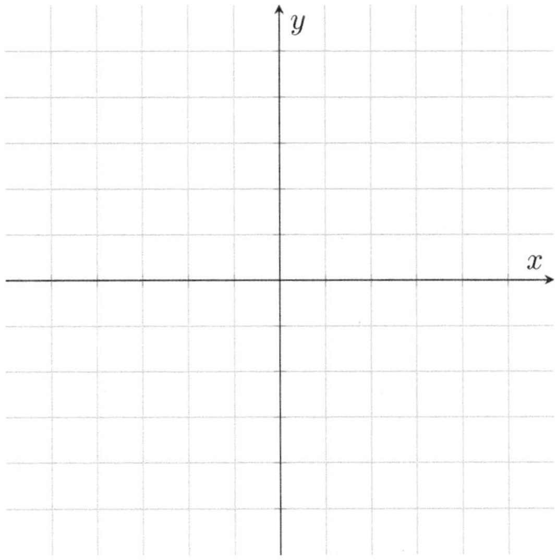

Example 5 (Graph system of inequalities)

$$\begin{aligned} 3x - y &\leq 6 \\ x + y &\geq 6 \\ y &\leq 2 \end{aligned}$$

	x-intercept	y-intercept	Test Point
Line 1			
Line 2			

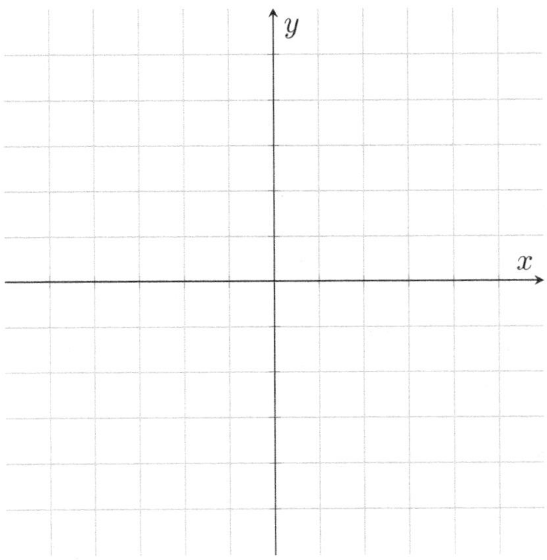

Example 6 (Automobile assembly plant)

An automobile assembly plant has an assembly line that produces the Hatchback Special and the Sportster. Each Hatchback requires 2.5 hours of assembly line time, and each Sportster requires 3.5 hours. The assembly line has a maximum operating time of 140 hours per week. Graph the number of cars of each type that can be produced in one week. Give three example solutions and one example that is not a solution.

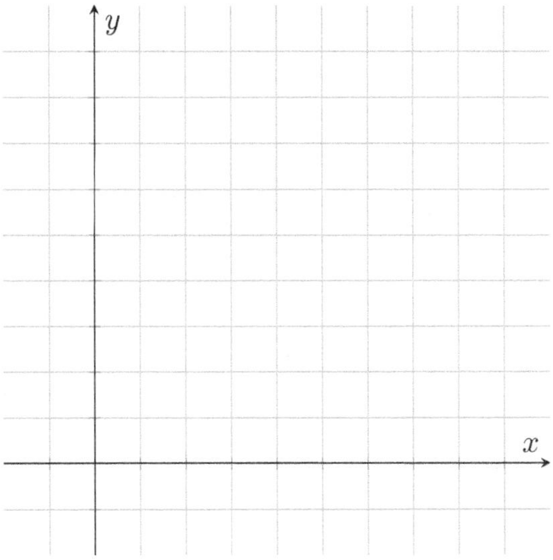

Example 7 (Write system of inequalities)

Write a system of inequalities that describes the graphed feasible region.

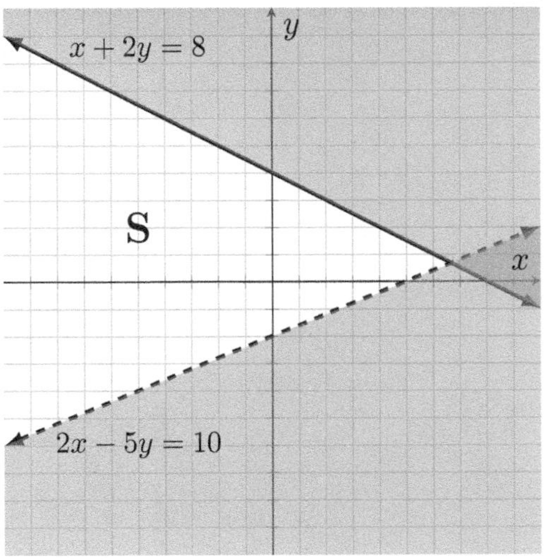

III. Corner Points

If the feasible region (or solution set) is completely enclosed by boundary lines, it is said to be **bounded**. If the feasible region is not enclosed, it is **unbounded**.

The intersection points of the boundary lines that are in the feasible region are called **corner points**. You may be able to identify some corner points by inspection. If you cannot identify the corner point(s), then you can use any convenient method from Unit 2 to find the point(s).

For Examples 8 - 9: Sketch the region that corresponds to the given inequalities. Label all lines and mark the feasible region with an S. State whether the feasible region is bounded or unbounded. List the corner points of the feasible region.

Example 8 (Sketch region and find corner points)

$8x + 16y \leq 32$
$x \geq 0,\ y \leq 0$

x-intercept	y-intercept	Test Point

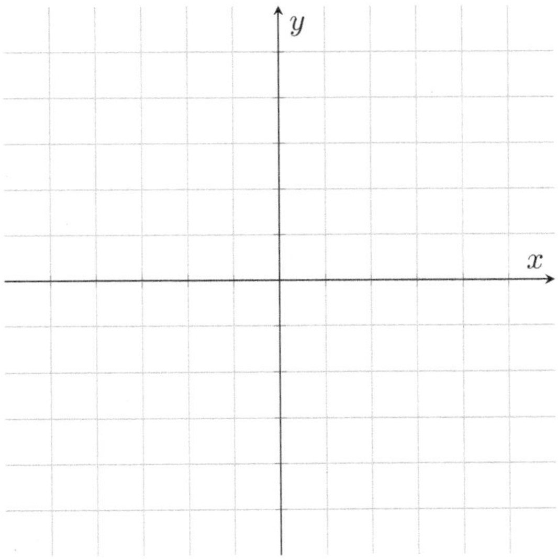

Example 9 (Sketch region and find corner points)

$$\begin{aligned} 2x + y &\leq 10 \\ x + y &\leq 8 \\ x \geq 0,\ y &\geq 0 \end{aligned}$$

	x-intercept	y-intercept	Test Point
Line 1			
Line 2			

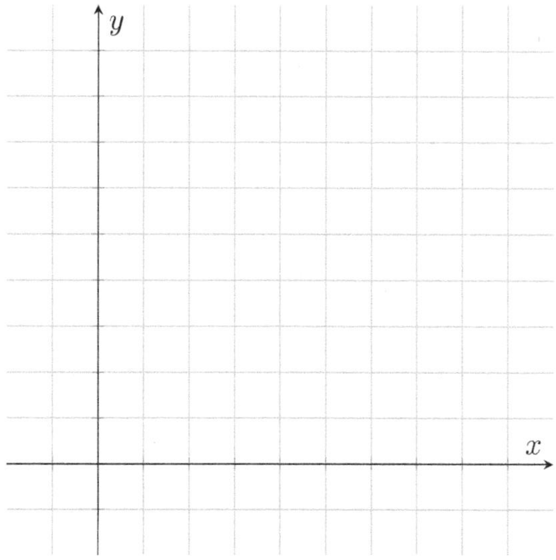

Example 10 (Investment strategy)

In 2003, the Fidelity Spartan Mass Fund (FDMXX) yielded 5% and the Fidelity Spartan Florida Fund (FFLIX) yielded 7%. You would like to invest a total of up to \$60,000 and earn at least \$3,500 in interest. Sketch the feasible region that shows how much money you can invest in each fund. Find the corner points of the region.

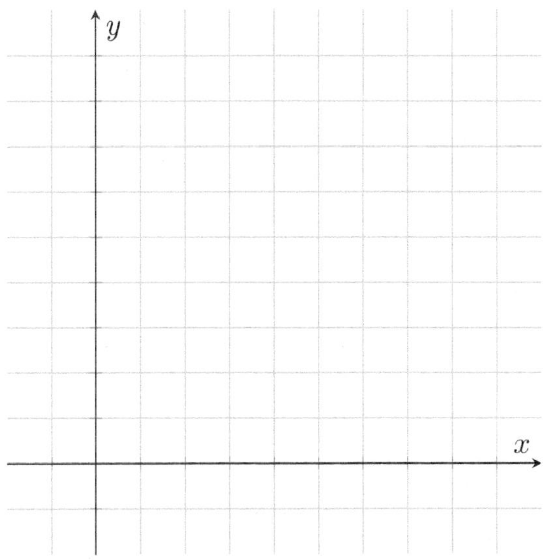

Additional Notes

NAME: ______________________ EMAIL: ________________ @vols.utk.edu

Lesson 12: Graphical Solution of Linear Programming Problems

Learning Objectives:

I. Determine whether a given problem is a linear programming problem.

II. Use the method of corners to solve a linear programming problem corresponding to a bounded feasible region.

III. Use the method of corners to solve a linear programming problem corresponding to an unbounded feasible region.

IV. Use the method of corners to solve a linear programming problem with no solution or infinite solutions.

V. Set up and solve an application problem using the method of corners.

I. Linear Programming Problems

A **linear programming (LP) problem** consists of a linear objective function to be optimized (i.e. maximized or minimized) subject to certain constraints in the form of linear equations or inequalities.

Example 1 (Determine if linear programming problem)

Determine whether each problem is a linear programming problem. If it is not, explain why.

(a.)

$$\begin{array}{ll} \text{Minimize:} & C = 2x + 3y \\ \text{subject to:} & 2x + 3y \leq 6 \\ & x - y = 0 \\ & x \geq 0,\ y \geq 0 \end{array}$$

(b.)

$$\begin{array}{ll} \text{Maximize:} & P = xy \\ \text{subject to:} & 2x + 3y \leq 12 \\ & 2x + y \leq 8 \\ & x \geq 0,\ y \geq 0 \end{array}$$

II. Bounded Regions

Theorem: Optimal Values

Given a linear programming problem, if the objective function has an optimal value (maximum or minimum), it must occur at a corner point of the feasible region.

The **Method of Corners** is a graphical procedure for solving linear programming problems based on the Optimal Value Theorem. This is a good method to use when we have a small number of variables and constraints.

To solve a linear programming problem using the method of corners:

1. Use each constraint to sketch the boundary of the feasible region.
2. Determine the corner points of the feasible region.
3. Evaluate the objective function at each corner point (i.e. plug each corner point into the objective function).
4. Find the corner point that meets the objective.

Theorem: Solution Existence (Part A)

If the feasible region is not empty and is bounded, then the objective function has both a maximum and a minimum.

For Examples 2 & 3: Solve the linear programming problem using the method of corners. Label all lines and mark the feasible region with an S when graphing.

Example 2 (Maximize and minimize - given graph)

Maximize and minimize $A = 6x + 15y$ subject to the feasible region shown in the graph.

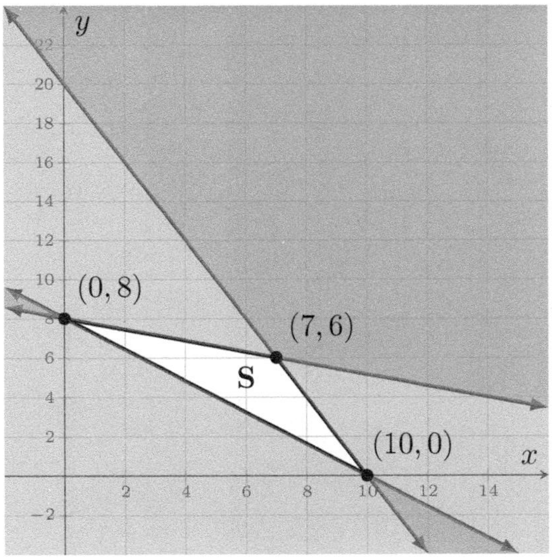

Example 3 (Maximize and minimize - not given graph)

Maximize and minimize $Z = 3x + 2y$ subject to $\begin{cases} x + y \leq 6 \\ x \leq 3 \\ x \geq 0,\ y \geq 0 \end{cases}$.

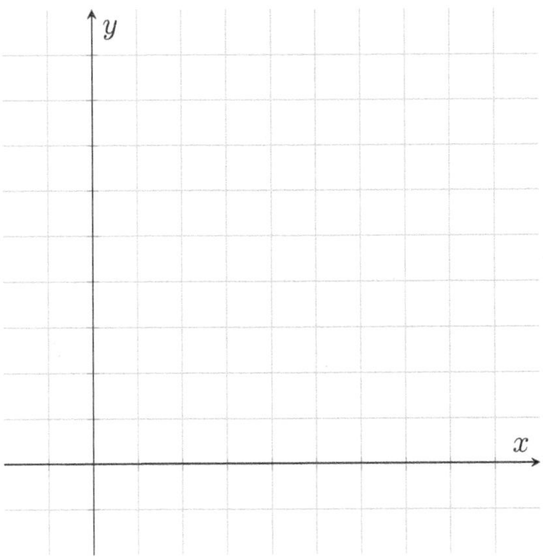

III. Unbounded Regions

Theorem: Solution Existence (Part B)

If the feasible region is unbounded and the coefficients of the objective function are non-negative, then the objective function has a minimum value (provided that the feasible region includes the non-negative inequalities). The objective function does not have a maximum value.

Example 4 (Unbounded region)

Maximize and minimize $C = 3x + y$ subject to $\begin{cases} x + 2y & \geq & 40 \\ x + y & \geq & 30 \\ x \geq 0, & y \geq 0 \end{cases}$ using the method of corners.

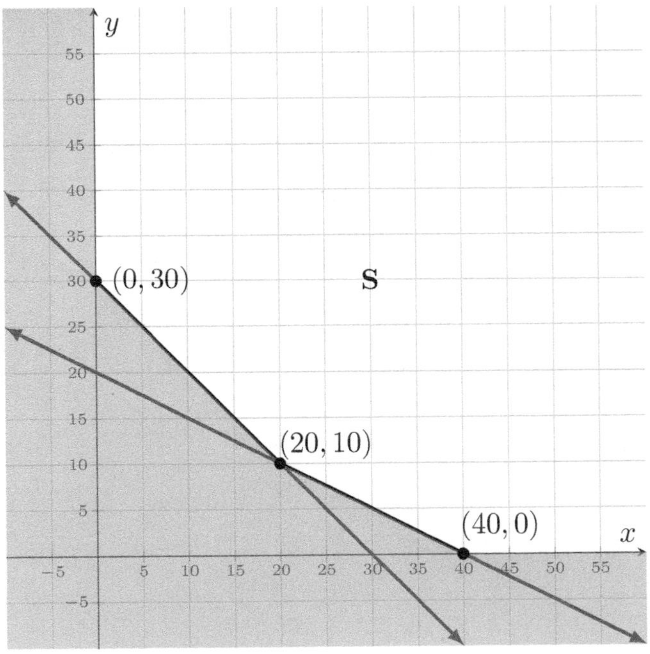

IV. Unusual Situations

> **Theorem: Solution Existence (Part C)**
>
> If the feasible region is the empty set, then the linear programming problem has no solution.

Example 5 (No solution - empty set)

Maximize $P = 9x + 13y$ subject to $\begin{cases} 3x - y \leq 8 \\ x - 2y \geq 5 \\ x \geq 0, \; y \geq 0 \end{cases}$ using the method of corners.

Label all lines and mark the feasible region with an S when graphing.

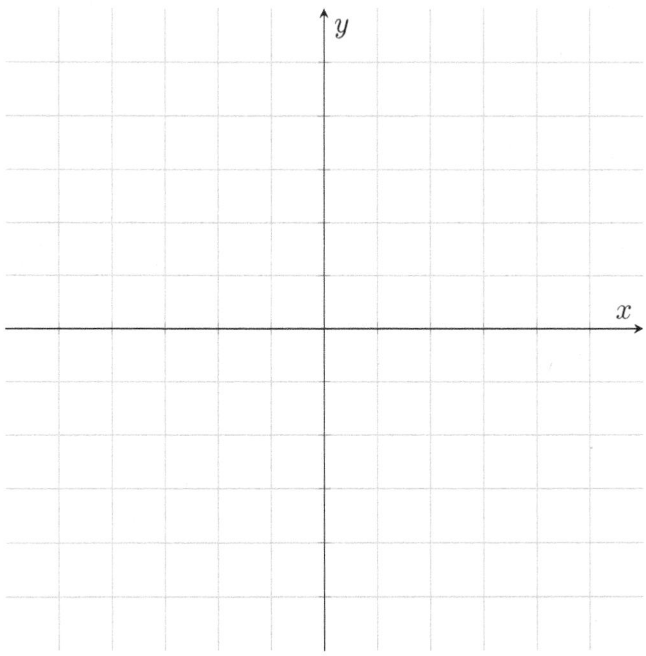

Theorem: Solution Existence (Part D)

If the objective function is optimized at two adjacent corner points of the feasible region, then it is optimized at every point on the line segment joining these points and therefore has infinitely many solutions to the problem.

Example 6 (Infinite solutions)

Minimize $C = 2x + 5y$ subject to the feasible region shown in the graph.

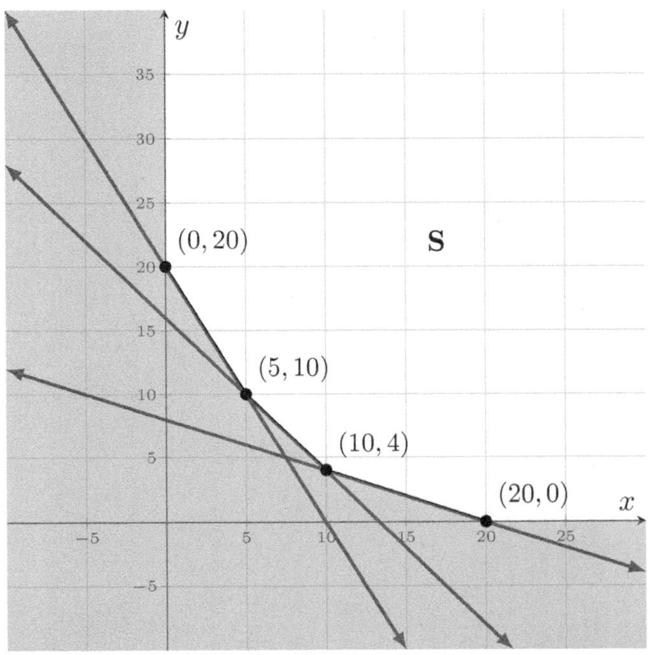

V. Applications

To formulate a linear programming problem:

1. Define the variables - look at the question for the problem to define **specific** variables.
2. State the objective - look at the question, state if it is a maximization or minimization problem, and state what the problem is optimizing.
3. Write the objective function - look at the details to write the equation for what is being optimized.
4. Write the constraints - look at the details of the problem to write the constraints.

Example 7 (Vegetable farmer)

A vegetable farmer has 75 acres of land and grows onions, carrots, and lettuce. It costs \$250 per acre to produce onions, and the profit is \$65 per acre. It costs \$300 per acre to produce carrots, and the profit is \$70 per acre. It costs \$325 per acre to produce lettuce, and the profit is \$50 per acre. Production costs cannot exceed \$225,000. Find the number of acres of each vegetable to maximize profit.

Define the variables, state the objective, write the objective function, and write the constraints. Do NOT solve.

Example 8 (National Business Machine)

National Business Machines manufactures two models of portable printers: A and B. Each model A printer costs \$100 to make, and each model B costs \$150. The profits are \$30 for each model A printer and \$40 for each model B printer. If the total number of portable printers demanded per month does not exceed 2,500 and the company has allocated no more than \$600,000 per month for manufacturing costs, how many units of each model should National Business Machine make each month to maximize its monthly profit?

Define the variables, state the objective, write the objective function, and write the constraints. Label all lines and mark the feasible region with an S.

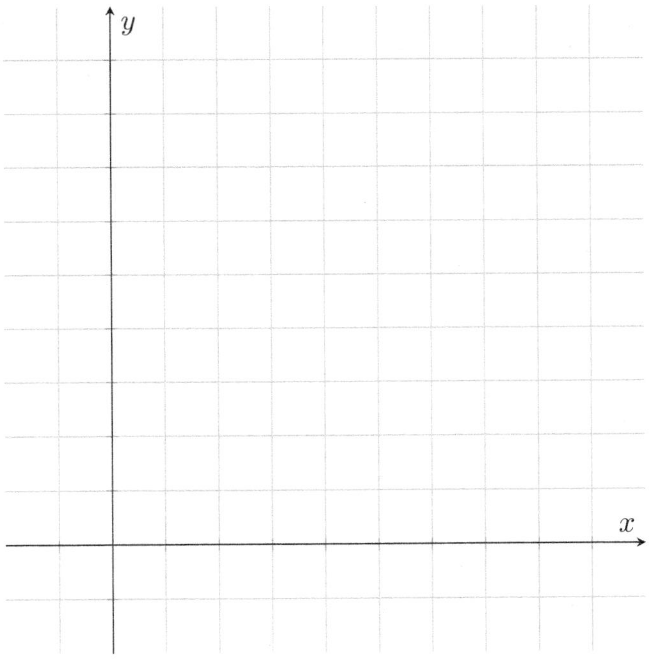

Summary of Lesson 12

Theorem: Solution Existence (Part A)

If the feasible region is not empty and is bounded, then the objective function has both a maximum and a minimum.

Theorem: Solution Existence (Part B)

If the feasible region is unbounded and the coefficients of the objective function are non-negative, then the objective function has a minimum value (provided that the feasible region includes the non-negative inequalities). The objective function does not have a maximum value.

Theorem: Solution Existence (Part C)

If the feasible region is the empty set, then the linear programming problem has no solution.

Theorem: Solution Existence (Part D)

If the objective function is optimized at two adjacent corner points of the feasible region, then it is optimized at every point on the line segment joining these points and therefore has infinitely many solutions to the problem.

Additional Notes

NAME: ______________________ EMAIL: ______________________ @vols.utk.edu

Lesson 13: Standard Maximization Linear Programming Problems

Learning Objectives:

I. Determine when a linear programming problem is a standard maximization problem.

II. Write slack variable equations, rewrite the objective function in standard form, and use these to set up a simplex tableau corresponding to a standard maximization problem.

III. Select the appropriate pivot column and row, and solve a standard maximization linear programming problem using the simplex method.

IV. Set up and solve a standard maximization application problem using the simplex method.

I. Standard Maximization Linear Programming Problems

Properties of a Standard Maximization Linear Programming Problem:

1. The objective function is to be maximized.
2. Each constraint is written using the $\leq$ inequality (excluding the non-negative constraints).
3. The constants in the constraints to the right of the $\leq$ are non-negative.
4. The variables are restricted to non-negative values (the non-negative constraints).

Example 1 (Determine if standard max LP problem)

Determine whether the following linear programming problems are standard maximization linear programming problems.

(a.)

$$\begin{aligned} \text{Maximize:} \quad & P = 6x + y + 3z \\ \text{subject to:} \quad & 3x + y \leq 15 \\ & 2x + 2y + 2z \leq 20 \\ & x \geq 0,\ y \geq 0,\ z \geq 0 \end{aligned}$$

(b.)

$$\begin{aligned} \text{Maximize:} \quad & P = 2x_1 + 3x_2 \\ \text{subject to:} \quad & -x_1 - 2x_2 \leq -12 \\ & 2x_1 + x_2 \leq 12 \\ & x_1 \geq 0,\ x_2 \geq 0 \end{aligned}$$

II. Setting up the Simplex Method

The **Simplex Method** is an algebraic technique used to solve linear programming problems. The process moves from corner point to corner point until an optimal solution is found. This process allows us to solve larger linear programming problems.

Refer to the following standard maximization linear programming problem to follow the steps for setting up the simplex method.

$$\begin{array}{ll} \text{Maximize:} & P = 3x + 5y \\ \text{subject to:} & x + 3y \leq 12 \\ & 2x + 3y \leq 18 \\ & x \geq 0,\ y \geq 0 \end{array}$$

Step 1: Introduce *slack variables* to turn the constraints into equations.

Step 2: Rewrite the objective function in *standard form*.

Step 3: Set up the *initial simplex tableau*.

Note: The total number of variables = (# of original variables) + (# of constraints) + 1 (for the objective function). This results in an underdetermined system.

Example 2 (Standard setup)

Express the linear programming problem as a system of equations using slack variables. Set up the initial simplex tableau.

$$\begin{array}{ll} \text{Maximize:} & Z = 6x_1 + x_2 + 3x_3 \\ \text{subject to:} & 3x_1 + x_2 \leq 15 \\ & 2x_1 + 2x_2 + 2x_3 \leq 20 \\ & x_1 \geq 0,\ x_2 \geq 0,\ x_3 \geq 0 \end{array}$$

III. Standard Simplex Method

Recall the standard maximization linear programming problem from Objective II.

Maximize: $P = 3x + 5y$
subject to:
$x + 3y \leq 12$
$2x + 3y \leq 18$
$x \geq 0,\ y \geq 0$

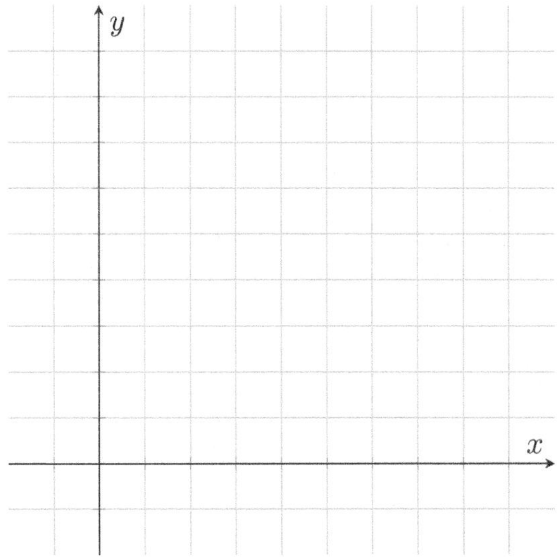

To get the **basic solution** corresponding to any simplex tableau:

- If a variable corresponds to a *unit column* in the simplex tableau, then the variable is an **active variable.** Its value is the constant in the row containing the 1.
- If a variable corresponds to a column that is *not a unit column* in the simplex tableau, then the variable is an **inactive variable.** Its value is 0.

Recall the initial simplex tableau from Step 3 which goes along with this standard maximization linear programming problem.

x	y	u	v	P	Constants
1	3	1	0	0	12
2	3	0	1	0	18
−3	−5	0	0	1	0

Step 4: Select the pivot column. Select the "most negative" coefficient from the bottom row. The column containing this entry will be the pivot column. *Step 4 ensures the greatest increase in the objective function.*

Step 5: Select the pivot row. Divide each constant above the horizontal line in the last column by the corresponding entries in the pivot column. The results are called test ratios. The row containing the smallest non-negative test ratio will be the pivot row. *Step 5 ensures we will not be taken from a feasible point to a non-feasible point.*

x	y	u	v	P	Constants
1	3	1	0	0	12
2	3	0	1	0	18
−3	−5	0	0	1	0

Step 6: Use a pivot program. Pivot the tableau on the selected element to move to the next basic feasible solution.

Step 7: Determine whether the optimal solution has been reached.

- If the last row contains any negative coefficients, then the objective function is not maximized. Repeat steps 4 - 6.
- If the last row **does not** contain any negative coefficients, then the final tableau has been reached. In this case, the basic solution is the optimal solution.

Example 2 (Standard simplex)

Finish solving the linear programming problem using the standard simplex method.

$$\begin{aligned} \text{Maximize:} \quad & Z = 6x_1 + x_2 + 3x_3 \\ \text{subject to:} \quad & 3x_1 + x_2 \leq 15 \\ & 2x_1 + 2x_2 + 2x_3 \leq 20 \\ & x_1 \geq 0,\ x_2 \geq 0,\ x_3 \geq 0 \end{aligned}$$

Initial Tableau:

x_1	x_2	x_3	s_1	s_2	Z	Constants
3	1	0	1	0	0	15
2	2	2	0	1	0	20
−6	−1	−3	0	0	1	0

For Examples 3 - 5: Determine whether the given simplex tableau is in final form. If so, find the optimal solution and optimal value to the associated linear programming problem. If not, find the pivot element to be used in the next iteration of the simplex method.

Example 3 (Optimal solution/value or next pivot)

x	y	u	v	P	Constants
0	$1/2$	1	$-1/2$	0	0
1	$1/2$	0	$1/2$	0	4
0	$-1/2$	0	$3/2$	1	12

Note: If a test ratio is 0, then we can only pivot on that row if the entry in the pivot column is positive. In other words, we can pivot on the row if the test ratio is $0/(\textit{positive } \#)$*, but not if the test ratio is* $0/(\textit{negative } \#)$*.*

Example 4 (Optimal solution/value or next pivot)

x	y	z	u	v	P	Constants
3	0	5	1	1	0	28
2	1	3	0	1	0	16
2	0	8	0	3	1	48

Example 5 (Optimal solution/value or next pivot)

x_1	x_2	x_3	s_1	s_2	s_3	P	Constants
3	1	1	$1/3$	0	0	0	28
2	2	0	$2/3$	1	0	0	16
2	5	0	$-1/3$	0	1	0	40
-2	-5	0	$5/2$	0	0	1	17

IV. Applications

Example 6 (Product manufacturing)

A company manufactures products: A, B, and C. Each product is processed in three departments: I, II, and III. The total available labor hours per week for Departments I, II, and III are 900, 1080, and 840, respectively. The time requirements (in hours per week) and profit per unit for each product are as follows.

	Product A	Product B	Product C
Dept. I	2	1	2
Dept. II	3	1	2
Dept. III	2	2	1
Profit	$18	$12	$15

How many units of each product could the company produce to maximize its profit? What is the largest profit the company can realize? Are there any resources left over?

Note: This linear programming problem actually has infinitely many solutions. In a later lesson, you will learn how to use the simplex method to identify infinitely many solutions.

Example 7 (Crop cultivation)

Interpret the final tableau given below for the following application problem.

A farmer has 150 acres of land suitable for cultivating crops A and B. The cost of cultivating crop A is $40/acre, whereas the cost of cultivating crop B is $60/acre. The farmer has a maximum of $7,400 available for land cultivation. Each acre of crop A requires 20 labor hours and each acre of crop B requires 25 labor hours. The farmer has a maximum of 3,300 labor hours available. He expects to make a profit of $150 per acre on crop A and $200 per acre on crop B.

How many acres of each crop should he plant to maximize his profit? What is the largest profit the farmer can realize? Are there any resources left over?

(Note: x = acres of crop A planted, y = acres of crop B planted, u = acreage slack, v = cost per acre slack, and w = labor hours slack.)

Final Tableau:

x	y	u	v	w	P	Constants
1	0	0	$-1/8$	$3/10$	0	65
0	0	1	$1/40$	$-1/10$	0	5
0	1	0	$1/10$	$-1/5$	0	80
0	0	0	$5/4$	5	1	25,750

Additional Notes

NAME: ______________________ EMAIL: ____________________ @vols.utk.edu

Lesson 14: Non-Standard Maximization Linear Programming Problems

---

Learning Objectives:

I. Determine whether a linear programming problem is a standard or non-standard maximization problem.

II. Set up and solve a non-standard maximization linear programming problem using the simplex method.

III. Set up and solve a non-standard maximization linear programming problem involving equality constraints.

I. Standard and Non-Standard LP Problems

Non-Standard linear programming problems are linear programming problems that cannot be written as standard maximization problems.

Example 1 (Standard or non-standard)

Identify whether the given problem is a standard maximization linear programming problem. State the solving method needed (either standard simplex or non-standard simplex).

(a.)

$$\begin{array}{ll} \text{Maximize:} & Z = 8x_1 + 5x_2 + x_3 \\ \text{subject to:} & x_1 + 6x_2 + x_3 \leq 0 \\ & 2x_1 + 4x_2 + 3x_3 \leq 10 \\ & x_1 \geq 0,\ x_2 \geq 0,\ x_3 \geq 0 \end{array}$$

(b.)

$$\begin{array}{ll} \text{Maximize:} & P = 2x + 3y \\ \text{subject to:} & -x - 2y \geq -12 \\ & 2x + y \leq 12 \\ & x \geq 0,\ y \geq 0 \end{array}$$

(c.)

$$\begin{array}{ll} \text{Maximize:} & P = 20x + 50y \\ \text{subject to:} & 3x + 4y \leq 72 \\ & 5x - 2y \geq 16 \\ & x \geq 0,\ y \geq 0 \end{array}$$

(d.)

$$\begin{array}{ll} \text{Maximize:} & Z = x - 2y \\ \text{subject to:} & 4x + 7y \leq 70 \\ & 2x + y = 20 \\ & x \geq 0,\ y \geq 0 \end{array}$$

II. Non-Standard Simplex Method

Solving with the Non-Standard Simplex Method:

- The difference in the simplex method for non-standard problems arises due to the negative constants in the last column.
- If there are negative entries in the constant column above the objective function row, then the current basic solution is not in the feasible region to begin with.
- Once all of the negatives in the last column are gone, we can continue with the standard simplex method.

Refer to the following non-standard maximization linear programming problem to follow the steps for setting up the non-standard simplex method.

Maximize: $P = 20x + 50y$
subject to: $3x + 4y \leq 72$
$5x - 2y \geq 16$
$x \geq 0,\ y \geq 0$

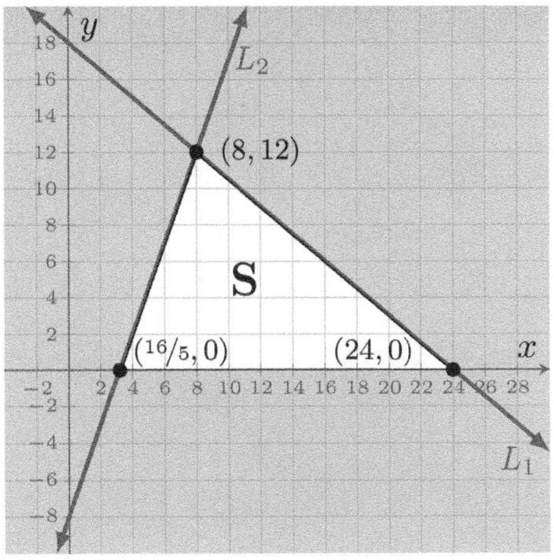

Step 1: Rewrite the problem. Rewrite the problem so that you can set up your constraints as slack variable equations. Write the objective function in standard form.

Step 2: Set up the initial tableau.

Step 3: Choose the non-standard pivot.

- Pick the "most negative" constant value from above the horizontal line. From this row, pick the column with the "most negative" coefficient. The column containing this entry will be the pivot column.
- The row containing the smallest non-negative test ratio will be the pivot row.
- Repeat this step until there are no negative entries in the last column (above the horizontal line). Then, continue solving using the standard simplex method.

Step 4: Get the optimal solution from the final tableau.

For Examples 2 & 3: Set up the slack variable equations, write the objective function in standard form, set up the initial simplex tableau, and circle the first pivot element. Do NOT solve.

Example 2 (Simplex setup)

Maximize: $P = 3x - 4y$

subject to:

$$\begin{aligned} x + y &\leq 45 \\ x - 2y &\geq 10 \\ x \geq 0,\ y &\geq 0 \end{aligned}$$

Example 3 (Simplex setup)

Maximize: $Z = x_1 + 3x_2 - 2x_3$

subject to:

$$\begin{aligned} 2x_1 + x_2 + 4x_3 &\leq 24 \\ x_1 + 2x_2 + x_3 &\leq 8 \\ 2x_1 - x_2 + 2x_3 &\leq 16 \\ x_1 \geq 0,\ x_2 \geq 0,\ x_3 &\geq 0 \end{aligned}$$

III. Equality Constraints

Example 4 (Equality constraints)

Solve the linear programming problem using the appropriate method.

$$\begin{array}{ll} \text{Maximize:} & Z = x - 2y \\ \text{subject to:} & 4x + 7y \leq 70 \\ & 2x + y = 20 \\ & x \geq 0,\ y \geq 0 \end{array}$$

Additional Notes

NAME: ______________________ EMAIL: ____________________ @vols.utk.edu

Lesson 15: Minimization Linear Programming Problems

Learning Objectives:

I. Solve a minimization linear programming problem by converting it to a maximization problem.

II. Set up and solve a minimization application problem using the simplex method.

I. Changing to a Maximization LP Problem

We solve minimization linear programming problems by converting them to maximization problems and using the methods from previous lessons to solve.

Example 1 (Minimization with $\leq$ constraints)

$$\begin{array}{ll} \text{Minimize:} & C = -2x + y \\ \text{subject to:} & x + 2y \leq 6 \\ & 3x + 2y \leq 12 \\ & x \geq 0,\ y \geq 0 \end{array}$$

Example 2 (Minimization with $\geq$ constraints)

$$
\begin{aligned}
\text{Minimize:} \quad & C = 2x + 3y \\
\text{subject to:} \quad & x + 4y \geq 8 \\
& x + y \geq 5 \\
& 2x + y \geq 7 \\
& x \geq 0, \ y \geq 0
\end{aligned}
$$

Example 3 (Minimization with = constraint)

$$\begin{array}{ll} \text{Minimize:} & Z = 3x_1 + 2x_2 + x_3 \\ \text{subject to:} & \begin{array}{rcl} x_1 + 2x_2 + x_3 & \leq & 20 \\ 3x_1 + x_2 & \leq & 30 \\ 2x_1 + x_2 + x_3 & = & 10 \end{array} \\ & x_1 \geq 0,\ x_2 \geq 0,\ x_3 \geq 0 \end{array}$$

II. Applications

Example 4 (Minimization application problem)

Fancy Pineapple produces pineapple juice and pineapple rings. This year the company anticipates a demand of at least 10,000 pints of pineapple juice and 1,000 cans of pineapple rings. Each pint of pineapple juice requires 2 pineapples, and each can of pineapple rings requires 1 pineapple. The company anticipates using at least 40,000 pineapples for these products. Each pint of pineapple juice costs the company 20 cents to produce, and each can of pineapple rings costs 50 cents to produce. How many pints of pineapple juice and cans of pineapple rings should Fancy Pineapple produce to meet the demand and minimize the total costs?

(a.) Define the variables, state the objective, write the objective function, and write the constraints.

(b.) Express the constraints as slack variable equations, write the objective function in standard form, set up the initial simplex tableau, and circle the first pivot.

(c.) Given the final tableau below, answer the question asked in the problem. Are there any extra resources?

x	y	u	v	w	P	Constants
1	0	0	$1/2$	$-1/2$	0	19,500
0	1	0	-1	0	0	1,000
0	0	1	$1/2$	$-1/2$	0	9,500
0	0	0	$2/5$	$1/10$	1	$-4,400$

Additional Notes

NAME: ______________________ EMAIL: ____________________ @vols.utk.edu

Lesson 16: Infinite Solutions and No Solutions

Learning Objectives:

I. Use the simplex method to solve a linear programming problem with infinite solutions.

II. Use the simplex method to solve a linear programming problem with no solution and identify whether the corresponding feasible region is unbounded or empty.

I. Infinite Solutions

We know from the graphical solutions that some linear programming problems have infinitely many solutions or no solutions (they don't all have one unique solution). What happens in the simplex method for each of these cases?

Determine Infinite Solutions from a Simplex Tableau

- If there is a zero in the bottom row of the final simplex tableau that is not a unit column, then the original linear programming problem may have infinitely many solutions.
- Find the optimal solution from the current tableau.
- To find another optimal solution, pivot in the non-unit column with the zero at the bottom and use the row with the smallest nonnegative test ratio.
- The new tableau gives the same optimal value at a different point. (Continue finding additional points as needed.)
- All points on the line segment joining the points are also solutions.

Example 1 (Infinite solutions)

Solve the linear programming problem with the method of corners and the simplex method.

$$\begin{aligned} \text{Maximize:} \quad & P = 18x + 24y \\ \text{subject to:} \quad & 3x + 4y \leq 48 \\ & x + 2y \leq 22 \\ & 3x + 2y \leq 42 \\ & x \geq 0,\ y \geq 0 \end{aligned}$$

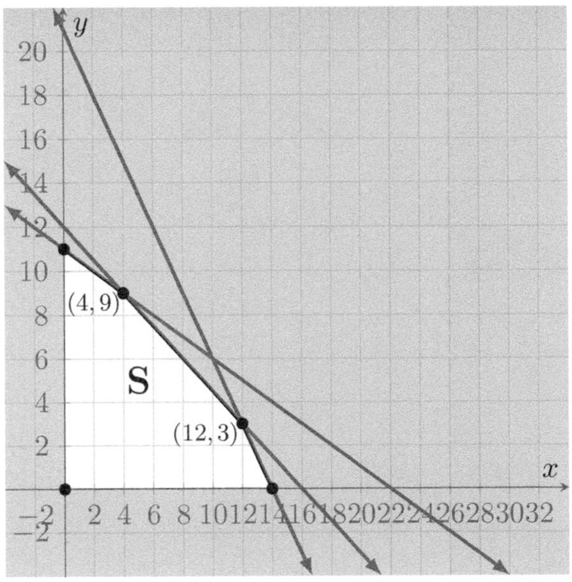

II. No Solutions

Two conditions give us the possibility of no solution in a linear programming problem.

1. No feasible region exists (i.e. the feasible region is empty).
2. The feasible region is unbounded, so the objective function has no maximum value.

Note: Whenever the final answer is "no solution," you must indicate the type of no solution (either no feasible region or unbounded).

Determine No Feasible Region from a Simplex Tableau

- If a simplex tableau has a negative entry in the last column and no other entries in the row are negative, then there is no feasible region.
- When there is no feasible region, the linear programming problem has no solution.

Example 2 (No solution - empty set)

Solve the linear programming problem with the method of corners and the simplex method.

Maximize: $P = 9x + 13y$
subject to: $3x - y \leq 8$
$x - 2y \geq 5$
$x \geq 0,\ y \geq 0$

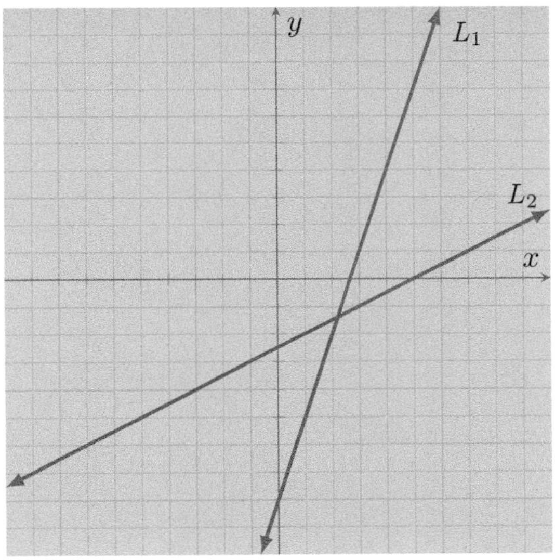

Determine Unbounded Solutions from a Simplex Tableau

- If a simplex tableau has no non-negative test ratios, then the feasible region is unbounded.
- There is no maximum value.

Example 3 (No solution - unbounded)

Solve the linear programming problem with the method of corners and the simplex method. Label your lines and mark the feasible region with an S.

Maximize: $P = 3x + y$
subject to: $x + y \geq 40$
$x + y \geq 30$
$x \geq 0,\ y \geq 0$

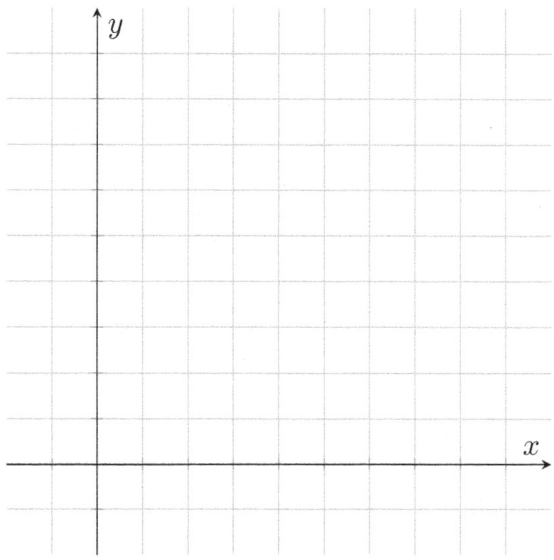

Additional Notes

Summary of Graphing Inequalities

To graph a linear inequality:

1. Draw the graph of the inequality as an equation (i.e. with an equal sign).
2. Use a solid line if the inequality is $\leq$ or $\geq$, and a dashed line if the inequality is $<$ or $>$.
3. Pick any test point **not** on the line.
4. Shade (or "block out") the side that does **NOT** contain solutions. Mark an S for the side that does contain the **solution set**.

To graph a *system of* linear inequalities:

1. For each inequality:
   (a) Draw the graph of the inequality as an equation (i.e. with an equal sign).
   (b) Use a solid line if the inequality is $\leq$ or $\geq$, and a dashed line if the inequality is $<$ or $>$.
   (c) Pick any test point **not** on the line and shade the appropriate region (i.e. the side that is false).
2. The region left empty is the **feasible region** or (**solution set**). Mark this region with an S.

Summary of Method of Corners

To solve a linear programming problem using the method of corners:

1. Use each constraint to sketch the boundary of the feasible region.
2. Determine the corner points of the feasible region.
3. Evaluate the objective function at each corner point (i.e. plug each corner point into the objective function).
4. Find the corner point that meets the objective.

Standard vs. Non-Standard Simplex Method

Do we have a standard maximization problem?

1. The objective function is to be maximized.
2. Each constraint is written using the $\leq$ inequality (excluding the non-negative constraints).
3. The constants in the constraints to the right of the $\leq$ are non-negative.
4. The variables are restricted to non-negative values (the non-negative constraints).

yes

Set up the slack variable equations, write the objective function in standard form, and set up the initial simplex tableau.

Are all the coefficients in the final row non-negative?

no

Pivot!

1. Select the "most negative" coefficient from the bottom row. The column containing this entry will be the pivot column.
2. Divide each constant above the horizontal line in the last column by the corresponding entries in the pivot column. The results are called test ratios. The row containing the smallest non-negative test ratio will be the pivot row.

yes

Solution

1. If a variable corresponds to a *unit column* in the simplex tableau, then its value is the constant in the row containing the 1.
2. If a variable corresponds to a column that is *not a unit column* in the simplex tableau, then its value is 0.

no, but maximization

If necessary, rewrite all constraints as $\leq$ inequalities.

Set up the slack variable equations, write the objective function in standard form, and set up the initial simplex tableau.

Are there any negative entries in the constant column above the horizontal line?

no

yes

Pivot!

1. Pick the "most negative" constant value from above the horizontal line. From this row, pick the column with the "most negative" coefficient. The column containing this entry will be the pivot column.
2. The row containing the smallest non-negative test ratio will be the pivot row.

Infinite & No Solutions

Determine Infinite Solutions from a Simplex Tableau

- If there is a zero in the bottom row of the final simplex tableau that is not a unit column, then the original linear programming problem may have infinitely many solutions.
- Find the optimal solution from the current tableau.
- To find another optimal solution, pivot in the non-unit column with the zero at the bottom and use the row with the smallest nonnegative test ratio.
- The new tableau gives the same optimal value at a different point. (Continue finding additional points as needed.)
- All points on the line segment joining the points are also solutions.

Determine No Feasible Region from a Simplex Tableau

- If a simplex tableau has a negative entry in the last column and no other entries in the row are negative, then there is no feasible region.
- When there is no feasible region, the linear programming problem has no solution.

Determine Unbounded Solutions from a Simplex Tableau

- If a simplex tableau has no non-negative test ratios, then the feasible region is unbounded.
- There is no maximum value.

Empty Graphs

The following empty graphs may be useful when sketching a graph on your own paper.

Remember that axes are assumed to count by 1's unless otherwise specified. When counting by anything other than 1's, the axes need to be labeled to indicate this.

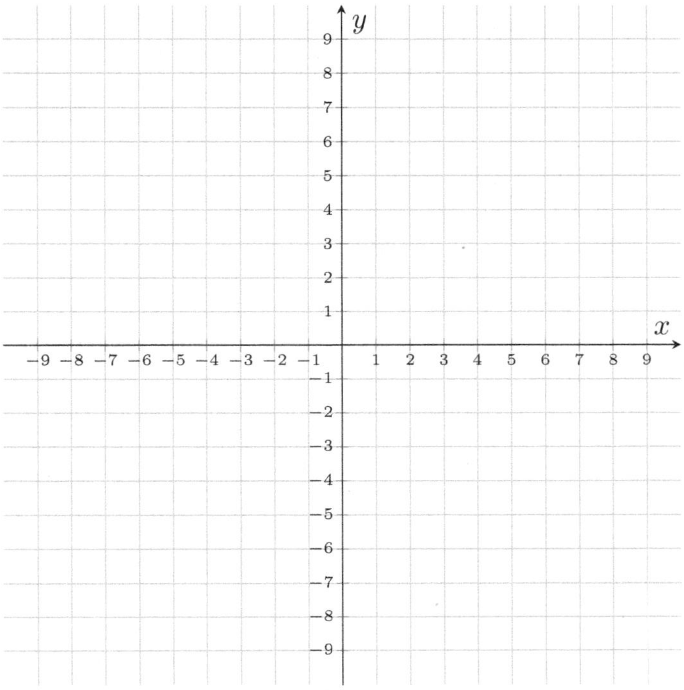

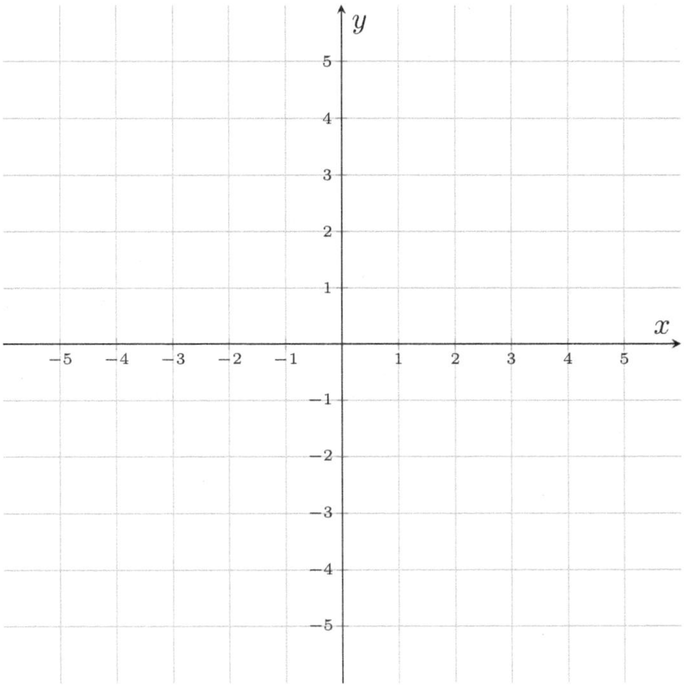

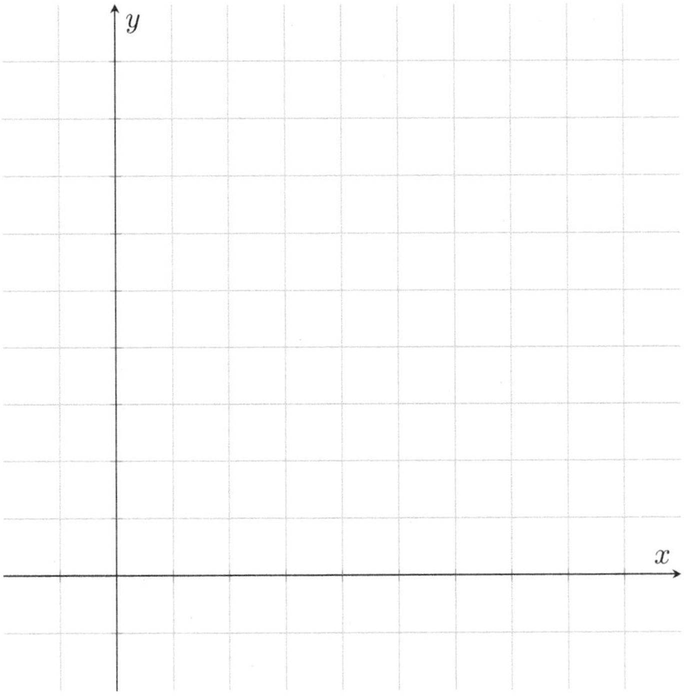

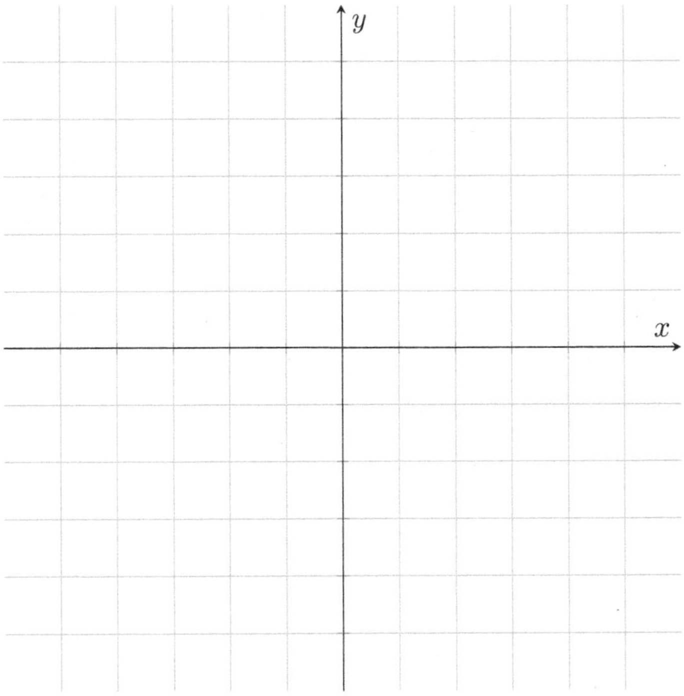

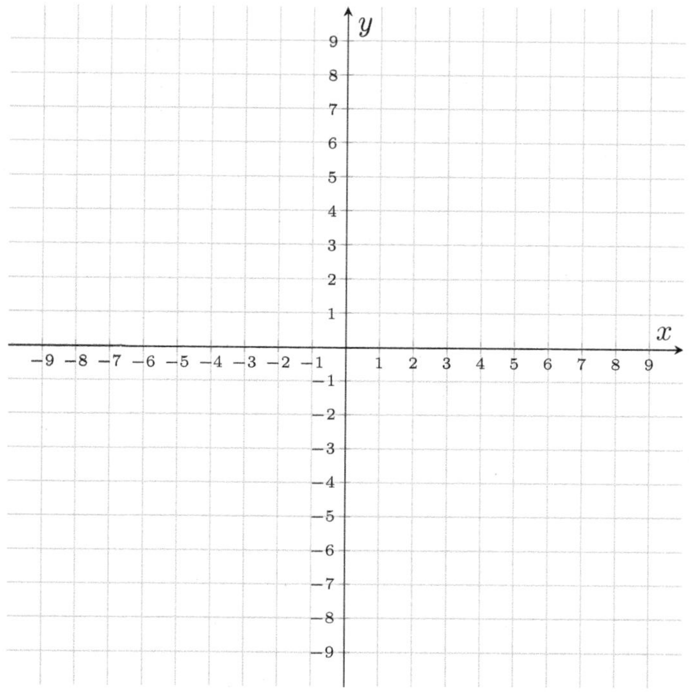
y
x
−9 −8 −7 −6 −5 −4 −3 −2 −1 1 2 3 4 5 6 7 8 9
9 8 7 6 5 4 3 2 1 −1 −2 −3 −4 −5 −6 −7 −8 −9

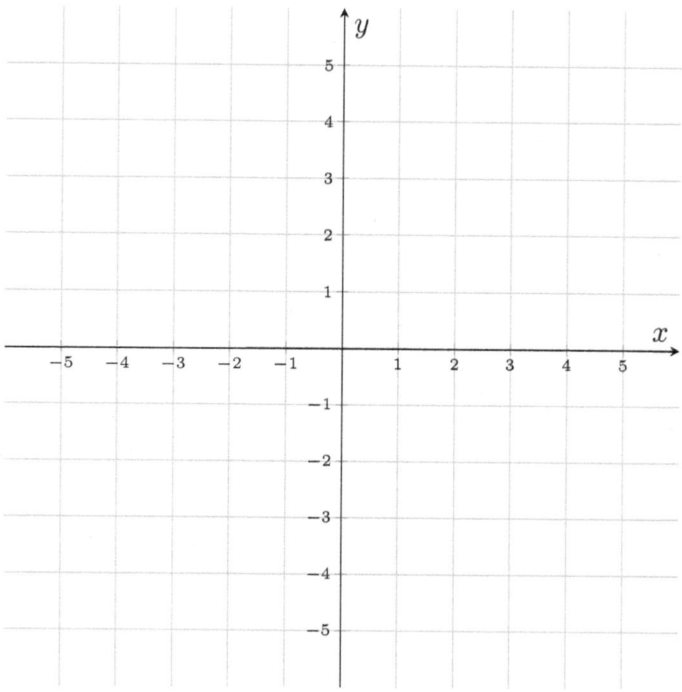
y
x
−5 −4 −3 −2 −1 1 2 3 4 5
5 4 3 2 1 −1 −2 −3 −4 −5

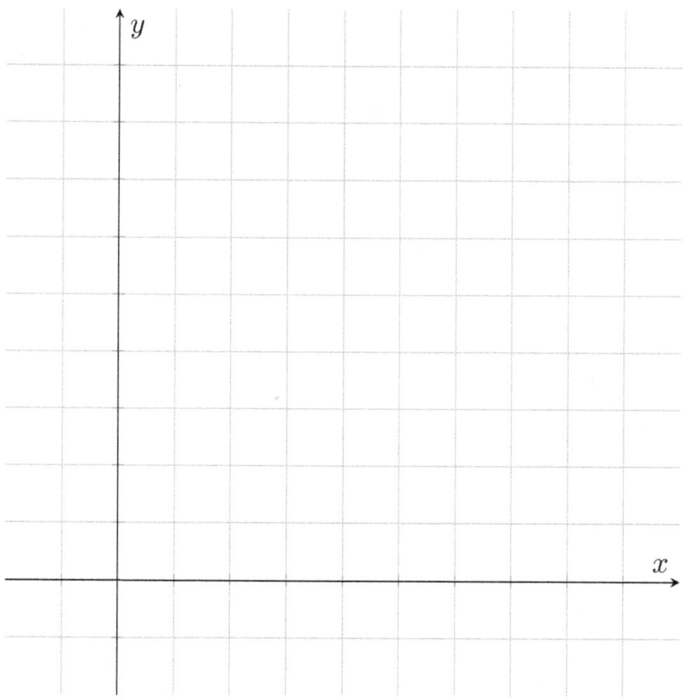
y
x

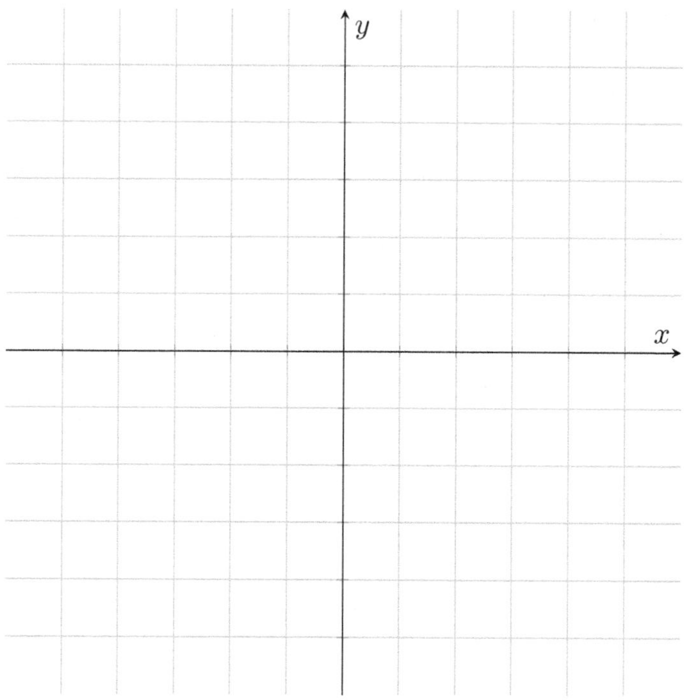
y
x

Unit 4

Game Theory

The theory of games is a relatively new branch of mathematics and owes much of its development to John von Neumann (1903–1957), one of the mathematical giants of the 20th century. John Harsanyi, John Nash, and Reinhard Selten won the Nobel Prize in Economics in 1994 for their work in this field.

This unit combines matrix methods with the theory of probability to determine the optimal strategies to be employed by two or more opponents involved in a competitive situation, with each opponent seeking to maximize their "gains," or, equivalently, to minimize their "losses." As such, the players may be poker players, managers of rival corporations seeking to extend their share of the market, campaign managers, or generals of opposing armies, to name a few.

NAME: ____________________ EMAIL: ____________________ @vols.utk.edu

Lesson 17: Game Theory and Strictly Determined Games

Learning Objectives:

I. Interpret or set up the payoff matrix corresponding to a two-person, zero-sum game.

II. Given a payoff matrix, compute the maximin strategy for the row player and the minimax strategy for the column player.

III. Determine whether a game is strictly determined; if so, find the saddle point, optimal strategy for each player, value of the game, and whether the game favors one player.

IV. Convert an application problem to a strictly determined game theory problem and solve it.

I. Game Theory

We will limit our study to a two-person, zero-sum game. A **two-person** game has exactly two sides (or players) involved. A **zero-sum game** is a game in which the two players compete for a payoff that one player pays to the other, so that the amount won by one player is exactly the same as the amount lost by their opponent.

We can represent the possible payoffs for the game in a **payoff matrix**: an $m \times n$ matrix in which there are m choices for the row player and n choices for the column player. The entries in this matrix are represented *from the row player's perspective.* The entries are positive if the play is a win for the row player and negative if it is a loss for the row player. The entry will be zero if the play is a tie.

Example 1 (Interpret payoff matrix)

Interpret the payoff matrix.

$$\begin{array}{c} \\ A \\ B \end{array} \begin{array}{c} \begin{array}{cc} C & D \end{array} \\ \begin{bmatrix} 3 & -4 \\ 0 & 1 \end{bmatrix} \end{array}$$

If the row player chooses move A and

- the column player chooses move C:

- the column player chooses move D:

If the row player chooses move B and

- the column player chooses move C:

- the column player chooses move D:

Example 2 (Set up payoff matrix)

You and your friend have come up with a game to pass the time. You simultaneously call out "heads" or "tails." If you both call heads, then your friend wins 2 points. If you both call tails, then your friend wins 1 point. If your calls differ, you get 2 points if you called heads and 1 point if you called tails. Set up the payoff matrix with you as the row player.

II. Pure Strategies

We call a player's plan of action against their opponent a **strategy**. Game theory attempts to determine the best strategy for each player. It assumes that each player knows all strategies available to all players, but each player selects a strategy without the opponent knowing. A player uses a **pure strategy** if they pick the same move every time when playing a game several times. The maximin and minimax strategies are both examples of pure strategies.

Maximin Strategy for the Row Player:

1. For each row of the payoff matrix, find the smallest entry in that row. (This is the *worst* outcome in each row for the row player.)
2. Choose the row for which the value found in Step 1 is as large as possible. (This is the *best* of the worst outcomes for the row player.)

(The row player's strategy is to choose a move where the smallest payout is as large as possible, so you are choosing the maximum of the minimums.)

Minimax Strategy for the Column Player:

1. For each column of the payoff matrix, find the largest entry in that column. (This is the *worst* outcome in each column for the column player.)
2. Choose the column for which the value found in Step 1 is as small as possible. (This is the *best* of the worst outcomes for the column player.)

(The column player's strategy is to choose a move where the largest payout is as small as possible, so you are choosing the minimum of the maximums.)

Note: For both the maximin and minimax strategies, we are choosing the "best of the worst."

Example 3 (Maximin and minimax strategies)

Determine the maximin and minimax strategies for the two-person zero-sum game represented by the following payoff matrices.

(a.)

$$\begin{bmatrix} 4 & -5 \\ -3 & -2 \end{bmatrix}$$

(b.)

$$\begin{bmatrix} 3 & 4 & -4 \\ 2 & -1 & -3 \end{bmatrix}$$

III. Optimal Strategies

The **optimal strategy** in a game is the strategy that is most profitable to a particular player. When playing a game several times, if the row or column player plays the same strategy every time, then the other player should be able to pick up on this and make an adjustment.

If no such adjustment can be made to either player's benefit when they figure out the play of the opponent's pure strategy, the game is said to be **strictly determined**. A strictly determined game is characterized by the following properties:

- There is an entry in the payoff matrix that is simultaneously the smallest entry in its row and the largest entry in its column. The location of this entry is called the **saddle point** for the game.

  Note: If the number you circled to the right when doing the maximin work is the same as the number you circled on the bottom when doing the minimax work, then the game is strictly determined. In this case, any entry where the maximin row(s) and the minimax column(s) intersect gives a saddle point.

- The optimal strategy for each player is their corresponding pure strategy, where the row player always plays a row corresponding to the maximin strategy and the column player always plays a column corresponding to the minimax strategy.

The entry in the saddle point of a strictly determined game is called the **value of the game**. If the value of the game is positive, then the game favors the row player. If it is negative, then the game favors the column player. If it is zero, then the game is called a **fair game**.

For Examples 4 - 6: Determine whether the two-person zero-sum game represented by the given payoff matrix is strictly determined. If the game is strictly determined, then:

(a.) Find the saddle point(s) of the game.

(b.) Find the optimal strategy for each player.

(c.) Find the value of the game.

(d.) Determine whether the game favors one player over the other. (If so, state which player is favored.)

Example 4 (Optimal strategy for payoff matrix A)

$$A = \begin{bmatrix} 3 & -3 & 0 & -1 \\ 1 & 2 & 1 & 0 \end{bmatrix}$$

Example 5 (Optimal strategy for payoff matrix B)

$$B = \begin{bmatrix} 1 & 2 \\ 0 & 3 \\ -1 & 2 \\ 2 & -2 \end{bmatrix}$$

Example 6 (Optimal strategy for payoff matrix C)

$$C = \begin{bmatrix} 2 & 4 & 2 \\ 0 & 3 & 0 \\ -1 & -2 & -1 \end{bmatrix}$$

IV. Applications

Example 7 (R-Mart and C-Mart)

Two major discount stores, R-Mart and C-Mart, compete for the business of the same customer base. R-Mart has 55% of the business, and C-Mart has the other 45%. Both companies are considering building new superstores to increase their market share. If both build, or neither build, they expect their market share to stay the same. If R-Mart builds and C-Mart doesn't, R-Mart's share increases to 65%. If C-Mart builds and R-Mart doesn't, then R-Mart's share drops to 50%. Determine which strategy (to build or not to build) each company should take.

Additional Notes

NAME: ______________________________ EMAIL: ________________________ @vols.utk.edu

Lesson 18: Games with Mixed Strategies

Learning Objectives:

I. Interpret or set up a mixed strategy matrix and compute the expected value of a game if the players use the given mixed strategies.

II. Solve a non-strictly determined game in which the payoff matrix is 2×2; write each player's optimal strategy as a row or column matrix.

III. Convert an application problem to a game theory problem and solve it.

I. Mixed Strategies

A player using a **mixed strategy** chooses each possible move a certain percentage of the time, but in a random pattern. The percentages for each move are written in fraction (or decimal) form, and should add up to 1.

Consider a game with an $m \times n$ payoff matrix:

$$A = \begin{bmatrix} a_{11} & \cdots & a_{1n} \\ \vdots & \ddots & \vdots \\ a_{m1} & \cdots & a_{mn} \end{bmatrix}$$

- The mixed strategy for the row player is represented by the row matrix:

  Note: p_i represents how often the row player plays row i.

- The mixed strategy for the column player is represented by the column matrix:

  Note: q_j represents how often the column player plays column j.

- The **expected value** of a game measures the average payoff to the row player when both players adopt a particular set of mixed strategies and play many times. The expected value of the game is given by the following formula.

Formula L. $E = PAQ$

Example 1 (Rachel & Carl)

Rachel and Carl play a game of matching fingers. On a predetermined signal, both players simultaneously extend two or three fingers from a closed fist. If the sum of the number of fingers is even, then Rachel receives an amount in dollars equal to that sum from Carl. If the sum of the number of fingers is odd, then Carl receives an amount in dollars equal to that sum from Rachel.

(a.) Set up the payoff matrix using Rachel as the row player.

(b.) Explain the following mixed strategies: $P = \begin{bmatrix} 1/4 & 3/4 \end{bmatrix}$ and $Q = \begin{bmatrix} 1/2 \\ 1/2 \end{bmatrix}$

(c.) Find the expected payoff with the given mixed strategies and explain its meaning.

Example 2 (Mixed strategy: 3×2 payoff matrix)

Consider a game with the following payoff matrix:

$$A = \begin{bmatrix} 1 & -2 \\ -1 & 2 \\ 3 & -3 \end{bmatrix}$$

(a.) Find the mixed strategy matrices for each player if the row player uses the maximin strategy 50% of the time and chooses each of the other two rows 25% of the time, and the column player chooses each column equally.

(b.) Find the expected value for the pair of mixed strategies described in part (a).

Example 3 (Mixed strategy: 2×2 payoff matrix)

Consider a game with the following payoff matrix:

$$A = \begin{bmatrix} 4 & -1 \\ -3 & 2 \end{bmatrix}$$

(a.) Find the expected payoff to the row player if both players use their pure strategy all of the time.

(b.) Find the expected payoff to the row player if the rower player plays row one 20% of the time and row two 80% of the time, and the column player plays column one $2/3$ of the time and column two $1/3$ of the time.

(c.) Which pair of mixed strategies is more advantageous to the row player? Explain.

II. Optimal Strategy

Given a 2×2 payoff matrix $A = \begin{bmatrix} a & b \\ c & d \end{bmatrix}$ for a game that is **NOT strictly determined**, we have the following formulas:

- The **optimal mixed strategy for the row player** is $P = \begin{bmatrix} p_1 & p_2 \end{bmatrix}$.

$$\textbf{Formula I.} \quad p_1 = \frac{d-c}{a+d-b-c}$$

- The **optimal mixed strategy for the column player** is $Q = \begin{bmatrix} q_1 \\ q_2 \end{bmatrix}$.

$$\textbf{Formula J.} \quad q_1 = \frac{d-b}{a+d-b-c}$$

- The **value of the game** is given by the **expected value** when the row and column players use their optimal mixed strategies, P and Q, respectively.

$$\textbf{Formula K.} \quad E = \frac{ad-bc}{a+d-b-c}$$

Example 4 (Optimal strategy: 2×2 payoff matrix, not strictly determined)

Find the optimal strategies for each player and the value of the game.

$$A = \begin{bmatrix} 2 & -1 \\ -3 & 2 \end{bmatrix}$$

Example 5 (Optimal strategy: 2×2 payoff matrix, strictly determined)

Find the optimal strategies for each player and the value of the game.

$$A = \begin{bmatrix} -1 & 3 \\ 2 & 5 \end{bmatrix}$$

III. Applications

Example 6 (Investment strategy)

The Smiths have decided to invest \$50,000 in the common stocks of two companies listed on the New York exchange. One of the companies derives its revenue mainly from its worldwide operation of hotel chains, whereas the other company is a domestic major brewery. It is expected that if the economy is in a state of growth, then the hotel stock should outperform the brewery stock; however, the brewery stock is expected to hold its own better than the hotel stock in a recessionary period. Suppose the following payoff matrix gives the expected percentage increase or decrease in the value of each investment for each state of the economy.

$$\begin{array}{r} \\ \text{Hotel} \\ \text{Brewery} \end{array} \begin{array}{c} \begin{array}{cc} \text{Growth} & \text{Recession} \end{array} \\ \begin{bmatrix} 25 & -10 \\ 10 & 15 \end{bmatrix} \end{array}$$

(a.) Determine the optimal strategy for the Smiths' investment of \$50,000.

(b.) What profit can the Smiths expect to make on their investment if they use their optimal investment strategy?

Additional Notes

NAME: ______________________ EMAIL: ______________________ @vols.utk.edu

Lesson 19: Game Theory and the Simplex Method

Learning Objectives:

I. Use linear programming to solve a non-strictly determined game in which the payoff matrix is larger than 2×2.

II. Determine the necessary method to solve a two-person, zero-sum game.

I. Optimal Strategy

We now begin the process of finding the optimal strategies for each player when the game is larger than 2×2 **and** not strictly determined. To do so, we use linear programming.

Example 1 (Linear programming with 2×3 payoff matrix)

Find the optimal strategies for each player and the value of the game.

$$A = \begin{bmatrix} 3 & -1 & 0 \\ -2 & 1 & -1 \end{bmatrix}$$

Step 1: Add a fixed number k to each entry of the payoff matrix so that all entries in the matrix become nonnegative AND no column is all zeros.

Step 2: Write ones to the right of and below the new matrix, with a zero in the bottom right corner. This is a matrix that we use to create a standard maximization linear programming problem.

Step 3: Solve the linear programming problem that was created in Step 2 by using the standard simplex method.

Step 4: Find the optimal strategies for each player and the expected value of the game using the following definitions/formulas.

- Row Player: Write the bottom row of the slack variables as a row matrix AND multiply each entry in the row matrix by $\frac{1}{Z}$.
- Column Player: Write the solution to the standard maximization problem as a column matrix AND multiply each entry in the column matrix by $\frac{1}{Z}$.
- Expected Value: Use the given formula.

Formula M. $E = \dfrac{1}{Z} - k$

Use the given final tableau to finish finding the optimal strategies and expected value.

x	y	z	u	v	Z	Constants
3	0	1	$3/5$	$-1/5$	0	$2/5$
-1	1	0	$-1/5$	$2/5$	0	$1/5$
1	0	0	$2/5$	$1/5$	1	$3/5$

Example 2 (Linear programming with 3×3 payoff matrix)

Find the optimal strategies for each player and the value of the game.

$$A = \begin{bmatrix} -1 & 1 & 2 \\ 2 & -1 & -2 \\ 1 & 2 & 0 \end{bmatrix}$$

(a.) Add $k =$ __________ to each entry. This gives us the new matrix:

(b.) State the associated standard maximization linear programming problem.

(c.) The final tableau is given below. Find the optimal mixed strategies for each player and the expected value.

x	y	z	u	v	w	Z	Constants
0	-3	0	$4/5$	1	$-8/5$	0	$1/5$
1	1	0	$-1/5$	0	$2/5$	0	$1/5$
0	$1/2$	1	$3/10$	0	$-1/10$	0	$1/5$
0	$1/2$	0	$1/10$	0	$3/10$	1	$2/5$

II. Identify Game Types

Example 3 (Identify game type)

Determine if the game associated with the given payoff matrix is strictly determined. State the method needed to find the optimal strategies (pure strategy, mixed strategy with formulas, or mixed strategy with linear programming).

(a.)

$$A = \begin{bmatrix} 1 & 2 & -3 \\ -1 & 2 & -2 \\ 2 & 3 & -4 \end{bmatrix}$$

(b.)

$$A = \begin{bmatrix} 2 & -1 & -2 & -3 \\ 1 & 2 & 0 & 1 \\ 0 & 2 & 3 & 3 \end{bmatrix}$$

(c.)

$$A = \begin{bmatrix} 2 & 5 \\ 4 & -1 \end{bmatrix}$$

Additional Notes

Unit 4 Additional Resources

How do we know which formula to use?

In Unit 4, one of our major goals is being able to determine the optimal strategy for each player and the value of the game. Formulas I - M allow us to complete these tasks. Again, even though the formulas are provided on the exams, it's important to know when to use each formula. Here's a guide that helps us decide which formula to use.

Formulas I, J & K

Formulas I, J, and K go together since they all deal with games that are **not strictly determined** and for which the payoff matrix is $\mathbf{2 \times 2}$.

- Use Formula I to find the **row player's optimal strategy**. This will tell us p_1, which is how often the row player should play row 1. Then compute $p_2 = 1 - p_1$. The row player's optimal strategy matrix is $P = \begin{bmatrix} p_1 & p_2 \end{bmatrix}$.
- Use Formula J to find the **column player's optimal strategy**. This will tell us q_1, which is how often the column player should play column 1. Then compute $q_2 = 1 - q_1$. The column player's optimal strategy matrix is $Q = \begin{bmatrix} q_1 \\ q_2 \end{bmatrix}$.
- Use Formula K to find the **value of the game.**

Formula L

Use Formula L to find the expected value of the game. We can use this formula whenever we are given (or we can find) mixed strategy matrices P and Q.

Formula M

Use Formula M to find the expected value of the game. We use Formula M whenever we have used linear programming to find the optimal strategies for each player. (Remember that we use linear programming when the game is not strictly determined AND for which the payoff matrix is larger than 2×2.)

Finding the Optimal Strategy

Given the payoff matrix for a two-person, zero-sum game, how do we find the optimal strategy for each player and the expected value of the game?
Start by doing the maximin and minimax work to determine whether the game is strictly determined.

The game is strictly determined.

- The row player's optimal strategy is to always play the row given by the maximin strategy.
- The column player's optimal strategy is to always play the column given by the minimax strategy.
- The expected value is the value in the matrix where the maximin and minimax strategies intersect.

The game is not strictly determined.

The payoff matrix is 2 × 2.

- The row player's optimal strategy matrix is

$$P = \begin{bmatrix} p_1 & p_2 \end{bmatrix},$$

where $p_1 = \dfrac{d-c}{a+d-b-c}$ and $p_2 = 1 - p_1$.

- The column player's optimal strategy matrix is

$$Q = \begin{bmatrix} q_1 \\ q_2 \end{bmatrix},$$

where $q_1 = \dfrac{d-b}{a+d-b-c}$ and $q_2 = 1 - q_1$.

- The expected value is $E = \dfrac{ad-bc}{a+d-b-c}$.

The payoff matrix is larger than 2 × 2.

- Add a fixed number k to each entry of the payoff matrix so that all entries are nonnegative and no column is all zeros.
- Write 1's to the right of and below the new matrix, with a 0 in the bottom right corner. Use this matrix to create a standard maximization linear programming problem.
- Solve the linear programming problem.
- To find the row player's optimal strategy matrix, write the bottom row of the slack variables as a row matrix and multiply each entry by $\frac{1}{Z}$.
- To find the column player's optimal strategy matrix, write the solution to the LP problem as a column matrix and multiply each entry by $\frac{1}{Z}$.
- The expected value is $E = \dfrac{1}{Z} - k$.